Natural Catastrophe

climate change and neoliberal governance

Brian Elliott

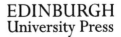

EDINBURGH
University Press

Edinburgh University Press is one of the leading university presses in the UK. We publish academic books and journals in our selected subject areas across the humanities and social sciences, combining cutting-edge scholarship with high editorial and production values to produce academic works of lasting importance. For more information visit our website: www.edinburghuniversitypress.com

Edinburgh University Press Ltd
The Tun – Holyrood Road,
12(2f) Jackson's Entry,
Edinburgh EH8 8PJ

Typeset in 11/13 Sabon by
IDSUK (Dataconnection) Ltd, and
printed and bound in Great Britain by
CPI Group (UK) Ltd, Croydon CR0 4YY

A CIP record for this book is available from the British Library

ISBN 978 1 4744 1048 9 (hardback)
ISBN 978 1 4744 1049 6 (paperback)
ISBN 978 1 4744 1050 2 (webready PDF)
ISBN 978 1 4744 1051 9 (epub)

Natural Catastrophe

Contents

Acknowledgments

I would like to acknowledge the many students who have taken my environmental ethics class at Portland State University (PSU) over the last five years. Their enthusiasm and commitment to environmental thought and action are constantly humbling and inspiring. The completion of this book was made possible by the award of a faculty enhancement grant from PSU. In addition, generous funding by the Institute of Sustainable Solutions at PSU allowed me travel to San Francisco to present at the Public Philosophy Network conference and to host an urban sustainability conference at PSU together with Shane Epting. I am grateful for many stimulating conversations with Steve Marotta on topics from contemporary ruins to modernist heroics. I would also like to thank Carol Macdonald, Michelle Houston, Nicola Ramsey and Ersev Ersoy at Edinburgh University Press. This book is twinned with our daughter, whose spirit will, I am sure, be equal to our future environmental challenges. I guess so, Felicity.

Introduction

THE THESIS OF THIS BOOK

The issue of climate change represents the cutting edge of global environmental concern. It is generally presented as a bewilderingly complex phenomenon, rigorously understood only by a small group of climate science experts. The impact of climate change, on the contrary, is universal and will rapidly become more severe as the present century progresses. This is especially the case in poorer parts of the developing world. The reality of climate change is generally acknowledged to amount to an unfolding environmental catastrophe. Its principal cause is the burning of fossil fuels, in particular coal, over the last two and a half centuries. Since the 1980s the political discourse that has arisen in response to the manifold environmental problems associated with industrialisation is centred on the idea of sustainable development (SD).

I will argue throughout this book that SD, while it is typically presented as politically neutral in character, offers in fact a very specific, neoliberal framing of environmental issues. Seen through the SD lens, climate change can only be tackled through radicalisation rather than abandonment of the global neoliberal order. On the other hand, many thinkers and activists currently tackling climate change are convinced that an ideological paradigm shift is needed to get to grips with the challenge of climate change. As Foucault (2008) insisted in his 1978/9 lecture series at the Collège de France, neoliberalism is not merely a matter of an ideology, that is, some sort of false consciousness giving agents an erroneous picture of material reality. Much more than a delusional state of consciousness, neoliberalism has become consolidated over the last four decades as a set of rigorous procedures and disciplinary practices governing individuals and institutions both from the 'inside' of personal consciousness and from the 'outside' in the form of social mores and political common sense. All too often, climate change is presented as an unintended consequence of capitalist development. Against this, I will argue here that both the creation of and inability to combat climate change are inherent features of neoliberal governance. Such a political order can no sooner move us beyond climate change than it can make good on its promises to eradicate global poverty.

To say this amounts to saying that climate change is not first an environmental concern and only secondarily and derivatively a political issue. Instead, climate change should be seen as a radically political matter. My basic argument in this book is that climate change is a characteristic and intrinsic symptom of neoliberalism. I use the term symptom here, rather than cause, in order to portray it as an aspect of neoliberal social pathology. The condition of neoliberal governance gives rise to this symptom but offers no genuine resources for the resolution of the problem of climate change. Marx's famous adage that humanity only sets itself problems it can solve can be understood in various ways. In a perverse way, this formula fits neoliberal governance very well. Neoliberal environmentalism is, clearly, marked by its enthusiasm for 'solutions'. My argument will be that the typical neoliberal solution to climate change conceals a broader tendency to undermine the potential for truly democratic social action. The general contours of the neoliberal solution to climate change are beginning to come into focus: a version of futures trading that will allow vast profits to be made against the backdrop of systemic ecological disturbance. While sound common sense presumes that no corporation or state could seriously consider climate change an entrepreneurial opportunity, the actual thinking of these agencies indicates otherwise (see Sullivan 2013). As will be discussed here in detail, climate change is already reshaping how the wealthiest states conceive of and deploy their military resources.

Neoliberal solutions to climate change, then, will inevitably perpetuate and deepen the kind of ecological consequences environmentalists take to be the real problem. From a properly environmental perspective, if we seek solutions to climate change these can only be found beyond the neoliberal horizon. But this is easier said than done. The initial difficulty takes the form of conceptualising forms of alternative political agency, principally a radical reconfiguration of the nation state relative to the thoroughgoing neoliberalisation to which it has been subject over the last four decades. The burgeoning literature on climate change, by contrast, largely takes for granted the model of neoliberal political agency and focuses on more specific means of action: public policy recommendations, grassroots activism programmes, or pointedly polemical stances on what climate change does or does not amount to. The heightened sense of urgency that tends to suffuse much of the literature on climate change militates against any thoroughgoing critique of neoliberal governance. We do not, it will be urged, have the time for such critique with the planet's ecological systems heading for collapse.

Against this sentiment, I will make the case here that this critique is in fact unavoidable if any plausible account of overcoming climate change is to be arrived at. This book is thus intended as an attempt to conceptualise an alternative paradigm of environmental political agency that would answer to the multiple social challenges inherent in climate change. It begins by presenting and critiquing the key features of the social imaginary of neoliberalism and then moves beyond this critique towards the construction of another kind of environmental agency identified in the work of critical geography. This is, therefore, not a book on technical solutions or policy recommendations, but rather a critical theory of environmental-political agency or a critique of political ecology.

OVERCOMING THE NATURE/POLITICS DIVISION

A useful way to begin this critique is with an extended reflection on the place of nature within politics. Here I am largely in agreement with Bruno Latour's (2004 [1999]) position, as set out his

Politics of Nature, that most environmental activism implies a concept of nature that is largely contradicted by the explicit pronouncements of environmentalists themselves. Latour situates environmentalism within a much broader conceptual scheme of modern politics according to which the natural and the political are considered mutually exclusive categories. This means that, according to standard environmental thinking, nature is nowhere to be found within politics as such. Nature is simply a term that refers to reality as studied by the natural sciences; whereas politics refers to the linguistic contestation between visions of the good society. Facts populate the first domain, values the latter. This split is replicated in the division of academic disciplines, according to which politics belongs to the humanities whereas nature is studied by the natural sciences.

In the public domain, the same split takes the form of a realm of politics subject to the largely irrational impact of public debate with its rhetorical and polemical forms of address (partisan politics), as opposed to another vision of policy-making led by the disinterested insights of scientists and technocrats (bureaucratic governance). Both sides, on Latour's view, are essentially wrong: politicians are not simply cyphers for irrational impulses on the part of the demos; and scientists are far from being socially impartial deliverers of facts. In actual fact, politics involves its own valid ways of knowing and science has its inherently political prejudices. Politics is not, as social contract theory would have it, the overcoming of nature; and the natural sciences are not predicated on an overcoming of partisan politics.

There is a further, more fundamental intertwining of politics and nature brought to light by Latour that is essential for the argument of this book. This constitutes another shibboleth of modernity, namely the claim that politics is essentially about the internal organisation of an exclusively human community. Latour insists, by contrast, that environmentalism implicitly rests on the insight that politics is (or should become) about the way all things (not just humans) interrelate. A veritably ecological politics works towards the democratic self-representation of non-humans: 'To limit the discussion to humans, their interests, their subjectivities, and their rights, will appear as strange a few years from now as having denied the right to vote of slaves, poor people or women' (Latour 2004: 69). Contemporary efforts to expand political protections enshrined in declarations of rights to non-humans are clearly indicative of this expanded sense of the political. One might capture this effort simply by saying it is about recognising non-human political subjects (as opposed to merely passive objects), but this invokes the very language of modernity that Latour seeks to move beyond. The argument is not that non-humans should, within environmentalism, be granted honorary human status, but that human agency can never be constituted without reference to the non-human. Latour is here pressing on some very fundamental, time-honoured aspects of western ontology, most obviously the uniqueness of human nature in virtue of which human beings alone count as ethical and political agents.

Equally, however, the earliest conceptualisations of nature that come down to us from ancient Greek philosophy suggest an ambiguity that Latour is arguably exploiting. These ancient writings on nature (in Greek, *phusis*) use the term to refer both to all things in their entirety (*kosmos*) and to the essential constitution of each type of thing (*essentia* as the Latin interpreters would come to call it). Latour is clearly leaning on the former term, whereas modernist politics accentuates the latter when it founds the political on the peculiar nature of the human being (following Aristotle in particular). Latour's argument is that modernity definitely settled this ambiguity in the concept of nature in favour of the latter meaning, thereby constituting politics as a function of an exclusively human mode of agency.

If we remain within the ambit of the modern conception, therefore, environmentalism can only ever amount to better or worse environmental management. Nature will refer to natural resources manipulated by human agents but cannot be considered constitutive of those very agents. If, by contrast, we follow Latour in emphasising the sense of nature that refers to all things in their intrinsic connectedness (the other sense of nature as *kosmos*), then environmentalism can no longer be seen as something human agents do to (or for) natural objects. It will rather capture the sense of co-organised agency involving humans and non-humans alike. While Latour explicitly distances himself from it, Marx's dialectical materialism can be useful for appreciating this point. In his early writings, most succinctly in his theses on Feuerbach, Marx (1994 [1888]) argued against the basic contention of Hegelian idealism that consciousness constituted the essence of human existence. Hegel denied that consciousness was itself a product of material processes and instead argued that it was the constituting source of all meaning relative to material objects, a prejudice that characterises all varieties of modern western rationalism and idealism from Descartes on.

NATURAL CATASTROPHE AND NEOLIBERAL GOVERNANCE

Marx views political-historical agency through the lens of capitalist crisis. Capitalism, with its inexorable drive to generate surplus production despite the barriers represented by local and national difference, is for Marx inherently crisis prone due to its internal 'contradictions'. The revolutionary class must look to the opportunities represented by such crises as pivotal moments to change the social and political order of things. Similarly, the predominant environmental presentation of climate change is one of systemic crisis or catastrophe. The key difference to be highlighted at the outset is that the social revolution Marx saw as the desired resolution to capitalist contradiction bears no comparison to the predominant way in which solutions to climate change are presented. This is because, for the most part, climate change is seen as a natural catastrophe. I will argue here, by contrast, that climate change is in essence a symptom of social pathologies that are political in origin. While the prevalent genre of climate catastrophism calls, ever more loudly, for solutions to what is seen as a complex natural phenomenon, the present analysis posits climate change as a problem of global politics. Making a case for this significant shift in perspective requires asking a second-order question: why has climate change arisen at this particular historical juncture as the key environmental question of our time?

Left-leaning environmental activism clearly borrows from Marxist crisis theory when it, not incorrectly, points to the patterns of capitalist globalisation as the root cause of climate change. But just as the long awaited final crisis of capitalism predicted by Marx has failed to turn up over the last century and a half, so has the environmental crisis thus far failed to generate any political agency of sufficient breadth and clout to challenge capitalist hegemony in plausible ways. Because, as initially indicated, climate change is predominantly conceived through a neoliberal lens, it is naive and unhelpful to conceptualise it as a natural catastrophe that will inevitably precipitate the kind of social revolution envisaged by Marx. To state only the most obvious reason for this: the legacy of the industrialised workers' movements has been all but liquidated across the wealthiest liberal democracies at the hands of neoliberalisation over the past forty years (see Srnicek and Williams 2015).

The phrase 'across wealthy liberal democracies' will appear regularly in this book. This indicates an intrinsic limitation of analysis that carries its own justification. It is often pointed out in discussions of climate change that the early-industrialising nations in the Global North, who are responsible for producing the vast majority of greenhouse gases over the last two and a half centuries, are not predicted to bear the brunt of the worst environmental effects. The environmental vulnerability of developing nations in the Global South is itself a complicated matter, involving geographic, demographic and social-political elements. The reason for the restricted scope of the present analysis is, in part, a matter of personal familiarity and conceptual competence. But it is also a reflection of the fact that global governance relative to climate change is predominantly shaped by the wealthiest nation states, most notably the United States. The role of such countries as China and India in climate politics largely remains to be seen and may plausibly induce some novel political consequences in the near future.

Because neoliberalism is the lens through which climate change is analysed here, it makes sense to restrict treatment to the most thoroughly neoliberalised political agencies and societies. This restriction in no way implies that environmental policy and activism beyond the Global North is unworthy of consideration. Quite to the contrary, many other studies make a convincing case that struggles for ecological sovereignty in the Global South have much to teach western-style political ecology (see de Castro, Hogenboom and Baud 2016). The United Nations Brundtland Report from 1987 (United Nations 1987) centred on an ethical claim to pursue economic development in such ways that equitable distribution of environmental benefits and burdens would be realised globally. While it is difficult to fault this aspect of SD, three decades on it is clear that the long-term effects of climate change will undoubtedly create a less, rather than more, environmentally equitable world.

Returning to the parallel between environmental and economic crisis, it is important to note that contemporary capitalism is quite different from that which was confronted by Marx in the mid-nineteenth century. Deindustrialisation within the leading world economies necessitated radical changes in social organisation. This included important transformations in how the working class viewed itself. On the level of popular political rhetoric these changes have led to a situation where any self-recognition of workers as a working class strikes an embarrassingly anachronistic note. The former industrial working class of the leading economies, long since splintered within a post-industrial economy dominated by the service sector with its attendant precariousness of employment, is relentlessly disciplined through a process of neoliberal restructuring (see Mason 2015; Srnicek and Williams 2015). Neoliberal 'govermentality', as Foucault referred to it, centres on a fundamental norm of self-entrepreneurship. Under this aspect, all personal development is grasped as autonomous self-exploitation. In other words, each individual operates within a social imaginary according to which self-improvement is the underlying maxim. Failure to realise potential in socially recognisable ways is at once a personal and social failing.

Take, for example, the neoliberalisation of higher education. Anyone who fails to take up tertiary education opportunities has made clear that they are not serious about improving their economic lot. This perspective leads, among other things, to a situation where discussions of the means and modes of education (for example, the model of education as social critique) become frivolous exercises only indulged in by antiquated and false radicals. The point for neoliberalism is not to ask what modes of higher education could be socially progressive but rather how to make such education a universally accessed social good as a marker of individual self-improvement. An intrinsic aspect of the neoliberalisation of higher education is, of course, making this sector gradually approximate to true market conditions (open consumer choice, economies of scale, coordinated competition and agglomeration, and the reduction or elimination of direct public subvention).

SUSTAINABLE DEVELOPMENT AND ENVIRONMENTAL POLITICS

A key initial question for us here is how climate change looks under the auspices of neoliberal governance. The preliminary answer to this can be encapsulated in one key word: sustainability. This term, gaining prominence in the 1980s and long since hegemonic in environmental discussions, first of all facilitates a mode of environmental activism that is inherently non-antagonistic with regard to global economic development. After all, sustainability began life as sustainable development. 'Development', in turn, meant and means greater integration into the global economy. Earlier environmental ideas, largely disseminated in the 1970s and proposing models of steady-state or even designed decline economies (see Meadows et al. 1972; Schumacher 1973), have been essentially effaced in the era of neoliberal governance. All this means that, after four decades of neoliberal social reconfiguration, any mention of environmental crisis will inevitably be translated into a variety of economic opportunity (see Funk 2015).

This amounts to what theoreticians refer to as 'weak' sustainability, namely the idea that loss of natural resources and habitat can be compensated for by equal or greater gains in technologically produced 'human capital'. Sustainability, I will argue in this book, is not simply compatible with the neoliberal conception of economic growth: it is a vital and core concept of neoliberalisation itself. It follows that attempts to use the paradigm of sustainability to create radical political critique are bound to fail. If my contention is correct, sustainability and the neoliberal management of natural resources amount to one and the same thing.

Those who, in theory or activism, are in earnest in their opposition to neoliberal governance should abandon any effort to render sustainability in more truly progressive forms. Environmental economics, in many ways the contemporary cutting edge of the sustainability paradigm, appears to offer the best hope that environmental concerns will be translated into effective public policy. Almost a decade ago Nicholas Stern (2007) published *The Economics of Climate Change*, an exhaustive account of the projected economic costs of climate change based on the most authoritative climate science then available. While Stern's methods of environmental accounting were criticised by others working in the field (see Nordhaus 2007), his work made clear that nation states would have to start taking seriously the financial implications of climate change in the very near, as opposed to distant, future. More broadly, the idea of 'full cost' or 'triple bottom-line' (economic, environmental, social/ cultural) accounting where all environmental impacts are factored into any development, has gained wide acceptance. And yet, for all the reams of data setting out the economic costs of climate change there have been no credible international efforts to rise to the challenge. At the present point in time, then, it is sufficiently clear that advances in climate science and economics cannot, of themselves, generate political praxis that would respond to the impending 'catastrophe' in credible ways.

More recently, a growing body of literature has grown up to account for this inaction. Psychologists, social, political and cognitive scientists have variously offered explanations for the apparent human inability to act on the evidence of systemic climate change (see Norgaard 2011; Marshall 2014; Hoffmann 2015). There are plausible elements in many of the theories offered but they mostly suffer from a strange aversion to political analysis in a more pointed sense. Typically, they rely on a version of what Aristotle first identified as weakness of will (*akrasia*), according to which an agent knows that something is in their best interest in principle but fails to act on that knowledge in a particular case. Hence, it is generally supposed that politicians are hobbled by corporate vested interest and individuals restricted to immediate rather than long-term self-interest (collective or individual ethical failure). Alternatively, states or individuals will only act if they can be sure others will act in congruent ways. Lacking any guarantees of reciprocity, all agents fail to act to further what they know to be the general good.

As Naomi Klein (2014) highlights in her recently published book *This Changes Everything*, climate change denial is a well-organised political force, particularly in the United States. Klein is convinced that this force sees the appeal to combat climate change as a political Trojan horse for ushering in a leftist political agenda opposed in principle to corporate America and its free trade vision of things. Klein offers a valuable short history of how many of the most respected environmental organisations in the United States have become more business friendly since the 1980s. She demonstrates the delight of the investigative reporter in revealing how non-profit organisations supposedly dedicated to the defence of nature have colluded in fossil fuel extraction. She remains surprisingly silent, however, on the harsh treatment meted out to more radical environmental action groups in the United States. Such groups include the Earth Liberation Front, who opposed logging in the Pacific Northwest through such actions as burning down facilities of the logging companies (see Vanderheiden 2005). Sidestepping the state's construction of 'eco-terrorism' in the wake of 9/11 (a surprising omission given her identification with the anti-globalisation movement of the 1990s), Klein instead identifies the way forward in the form of groups such as 350.org, which stages non-violent protest internationally to raise awareness of the immediate and long-term dangers of climate change.

More significantly, Klein also downplays the likelihood that casting climate change as a crisis calling for immediate and quite draconian action will almost certainly lead to a strengthening rather than weakening of the stranglehold corporations have on the political agency of the nation state. This is all the more surprising given that her previous book, *The Shock Doctrine* (Klein 2007), precisely highlighted how neoliberalism thrives by taking advantage of the social confusion sown by real or projected crisis. It is curious that Klein seems largely unaware of this eminent potentiality connected to framing climate change as natural catastrophe. Klein's vision of environmental social change adopts the approach of most accounts of western progressives, namely the hope that non-violent action in the form of organised popular protest will gain sufficient momentum to overcome a sclerotic state apparatus and the machinations of global corporations. Given that the stakes are so high and the timeline so short, this hope really amounts to wishful thinking in the absence of a coherent alternative social imaginary.

Rather than grasping climate change as a short-range catastrophe it should, as I argue throughout this book, be seen as a call to start the painstakingly slow process of reassembling the kind of critical popular politics that characterised various socialist traditions from the mid-nineteenth to mid-twentieth centuries. This will, of course, amount to another kind of socialism. Mainstream progressive environmentalism as it currently stands seems to believe, by contrast, that the economic and political drivers of climate change will be somehow undermined simply by getting enough people out onto the streets in protest. The consolidated position of the neoliberal political class, however, shows itself time and again largely indifferent to such protest. The post-ideological condition that has increasingly characterised western politics since the 1980s has to date not been disturbed, and has arguably been bolstered, by an environmental movement that habitually depoliticises its cause by appealing to 'the planet' as some sort of unproblematic common concern.

An additional worry I have about Klein's analysis relates to her celebration of localised environmental activism. Environmental localism is a complicated phenomenon and I will return to it repeatedly throughout this book. Provisionally it can be said that environmental localism largely unwittingly supports the neoliberal undermining of the nation state as a self-standing political agent. In supplanting Keynesian economics neoliberalism directly challenged the state's legitimacy on such counts as striving for full employment and maintaining a comprehensive welfare state. Environmental localism perpetuates the neoliberal critique of the nation state, thereby making credible climate action significantly less likely. A key premise of my argument here will be that progressive environmentalism cannot abandon the state as environmental agent and instead look to civil society as the sole source of authentic ecological self-determination (see Srnicek and Williams 2015: 25–49).

RESOURCE SECURITY AND ENVIRONMENTAL LOCALISM

To insist on the positive role that must be played by the nation state in response to climate change is not to deny the predominantly miserable record of environmental public policy across liberal democracies in recent decades. Viewed through the lens of sustainability and its focus on negative economic impacts, environmental crisis largely serves to justify an ever more stringent regulation of civic life in the name of state security. In the post-9/11 era the general public across liberal democracies have become used to increased surveillance and onerous security procedures in the context of international travel. While oil extraction still largely fuels the world, new extractive technologies have allowed richer countries to move towards a long-cherished goal of energy independence. In the age of climate change relying on the fossil fuel resources of other nations has become a key political concern. For many citizens of wealthier nations it has been profoundly shocking to see fossil fuel extraction on their own doorstep.

In this context the greatest headline-grabber by far has been hydraulic fracturing or 'fracking' to obtain natural gas. The tar sands in Alberta, Canada, represent a further key instance of this push towards domestic fossil extraction. Scientific studies of these new means of energy production make clear that they carry a very high environmental cost (see Wood et al. 2011). They represent a serious setback to hopes that the wealthiest nations would finally start to make serious investments in renewable energy. But the environmental costs of these new methods of extracting fossil fuels are more than cancelled out in the eyes of many politicians by the political gains. Appealing to a populist vision of national energy independence plays very well rhetorically in an era of neoliberal post-industrialisation. The same rhetoric can and often is indulged in when promoting domestic green jobs. And yet, as the western solar electricity industry has experienced in recent years, the same rules of global free trade that generally make Indian call centre workers much cheaper than their first world counterparts ensure that China will always be able to produce cheaper solar panels (see Lewis 2014). If national politicians were in earnest about energy independence they would have to start dismantling the international rules of trade that render largely illegal any sheltering of domestic economies against the free global trade in commodities.

Clearly, desire for control over energy resources prompts the neoliberal state to all manner of domestic and foreign intervention. In the shadow of climate change, this is likely to become more rather than less the case with regard to environmental resources such as agricultural land and water. Agriculture has long since been transformed into a massive trading network of food commodities. As climate patterns become less predictable the consequent fluctuations in agricultural yields will inevitably prompt powerful states to devote more resources to guaranteeing national food security. On the other hand there is a clear mismatch of scale between the state security agenda and environmental activism, which largely prefers to act at the local or regional scale. This preference is often justified by a pervasive sense that only at ground level can the authentic relationships be built that will allow us to care for natural resources in genuinely insightful ways. This is nowhere more obvious that in the burgeoning local food movements that have arisen in the wealthiest nation states over the last two decades.

The problem for advocates of environmental localism more generally, however, is that there are, in fact, no guarantees that local residents will act in a more informed or benevolent fashion with respect to their environmental resources than outsiders. As the case histories of environmental justice make clear, the ability of a community to make savvy environmental decisions largely depends on the economic and participatory power of the community in question. In an economically marginalised town or neighbourhood raw need is more likely to make residents acquiesce with environmentally harmful agencies even when they know the threats to their long-term interests. We need only keep the same tendencies in mind to see the limits of environmental localism when it comes to climate change. As I write, Pacific island nations are preparing strongly worded appeals to the most economically powerful states at the Paris climate negotiations of November 2015. Even where these most vulnerable communities demonstrate remarkable environmental consciousness, their localised agency is clearly incapable of action sufficient to restrict the rising oceans already inundating their territory.

At the other end of the scale of political agency, the recent history of intergovernmental climate talks makes clear that here too prospects are not good for slowing down climate change. At one international meeting after another states have generally operated, singly or in blocs, according to the dictates of comparative economic advantage. While the long history of the causes of climate change undoubtedly complicates international negotiations, it is the model of neoliberal governance that primarily accounts for the lack of effective international environmental policy and action. Thus, the best way forward lies in undermining the neoliberal configuration of state agency. Widespread popular disenchantment with the state and its political class has long since become a media cliché. Diagnosing the causes of this disenchantment has given rise to a cottage industry of popular pundits and academics alike. My sense is that the cause of the disaffection is actually fairly straightforward and essentially a product of the eclipse of a popular workers' movement across all contemporary liberal democracies.

Many indices of this eclipse can be cited, such as the drastic decrease in union membership, exacerbated economic inequality, or the increased influence of corporate money over the political class. Against this backdrop it is vital to appreciate that neoliberalism is a very specific, indeed unique variant of global capitalism. It has flourished in the precise circumstances of a supposedly post-ideological global condition following the collapse of Soviet state communism. While the origins of neoliberalism as a specific conceptualisation of state action lie back in the first half of the twentieth century and came to a certain conceptual fruition with the post Second World War reconstruction of Europe, it was only after the demise of the Soviet bloc that neoliberal governance could assume virtually unchallenged status as global political common sense. Despite waves of anti-globalisation and anti-capitalist struggles over the last two decades, there are few if any cogent signs that neoliberalisation is weakening. What is happening to public education in the United States, the United Kingdom and elsewhere is testimony to the fact that there is no obvious end in sight for the neoliberal reconstitution of society in general and the public sphere in particular.

With this in mind, it is difficult to lend credence to those accounts of climate change (exemplified by Klein) that herald sudden, progressive social change given the environmental dangers we face. On the contrary, as I shall argue throughout this book, there is much more evidence that climate change is being embraced by corporations and neoliberal national governments as the big security issue of the immediate future. In other words, climate change is being constituted as an object of concern within the all-pervasive neoliberal horizon. We have long since grown used to the rhetoric of energy security, an agenda particularly pursued by the United States in recent decades. It takes no great leap of imagination to appreciate how the concern for energy security can be readily extended to such things such as food supply, which requires land, water, and naturally occurring fertiliser inputs. While energy independence rhetoric appeals to domestic electorates that nostalgically yearn for an earlier era when western economies were industrial powerhouses, evoking environmental self-determination in the face of global climate change is not so easy. Even without a concrete sense of the science behind it, most people readily recognise that climate change is a global phenomenon that does not respect national (let alone bioregional) boundaries.

Similarly, the nature of intensified commodity agriculture means that it can never be in the interest of neoliberal nation states to pursue an agenda of agricultural self-sufficiency. Even an economic bloc such as the European Union, which enjoys great environmental diversity and capacity to produce most food products European consumers want, is a massive importer of food (see European Commission 2016). The widely cited negative externalities of industrial agriculture are scarcely more factored into the costs of capitalist production than in any other sector. Thus, it makes economic sense for a nation or bloc of nations to import food even when they could produce it themselves, albeit at a higher price. Indeed, in reality the expanded European Union has exacerbated internal agricultural competition, thereby undermining its own professed efforts to support localised agricultural diversity on environmental, cultural, and historical grounds (see Steen-Olsen et al. 2012).

At a global level, an intensified 'land-grab' phenomenon, associated with rapid increases in the cost of agricultural land, has become relatively well known (see Cotula 2013). The policy documents of the United Nations that relate to food are predicated on a pervasive sense of food as primarily a security issue. The prediction that global population will peak at around 10 or 11 billion by mid-century is usually offered as a fundamental fact on which this security rhetoric is based. What goes largely unquestioned in UN policy documents is the environmental desirability of the main drivers of population increase, among which rapid urbanisation is key. Urbanisation is inevitably paired with depopulation of rural areas, as the increased urban population must initially come from elsewhere. But this is, in turn, driven by a political first principle according to which a lower percentage of a nation's population working the land is a proxy for higher economic development. The ideal, as exemplified by western nations, is to have no more than 2 or 3 per cent of the population in the agricultural sector.

Desertion of agricultural lands clears the way for intensive patterns of food production dominated by the massively consolidated global agro-business sector. This sector, following the pattern of globalisation more generally, does not recognise any claims of the native population to be fed, but simply looks for the most lucrative ways to sell its products. Food aid delivered by benevolent western states to the malnourished residents of developing countries is thus, in great part, a phenomenon of neoliberal agricultural exploitation (see Clapp 2006, 2012). Even the wealthiest nations have grown to accept food aid as a domestic reality, with a large segment of the federal budget allocated to the United States Department of Agriculture going to food stamps. This is augmented by ubiquitous food banks, most recently appearing in the United Kingdom in the wake of the Great Recession, where politicians have lauded them as manifestations of the 'big society' (that is, neoliberal divestment of the welfare state in favour of a 'third sector' provision of social goods) in action (see Gentleman 2014).

The idea that public money should add to the profits of large corporations while those most economically marginalised are expected to resort to mutual aid is a hallmark of neoliberalisation. The citizens, it seems, must help themselves; big business, by contrast, is always first in line for a government handout. Such are the bizarre realities of the neoliberalised state. Meanwhile, even where localised food production remains relatively strong among developing nations every effort is made to frustrate unmediated access to western markets. This has given rise to the Fair Trade movement, which in reality gives back to producers only a tiny fraction of the western consumer price (see Jaffee 2014). Better than nothing no doubt, but hardly a challenge to the fundamental unfairness that undergirds putatively free global agricultural markets.

NATURE AND POLITICS

The extension of the national security agenda to straightforwardly environmental issues is a clear current tendency and as such a key salient feature of contemporary neoliberal governance. The same logic that is deployed when western corporations buy the right to exploit the fossil fuel reserves of developing nations applies in cases of environmental security. While there has been concerted and courageous local opposition to such environmental exploitation, the nation states affected, in many cases heavily burdened by a post-colonial legacy, have mostly given way to external economic and political forces at the expense of the conditions of their own people (see Umejesi and Akpan 2013). It is not hard to recognise here a familiar pattern of neoliberal globalisation, whereby wealthier nations exploit poorer nations and then blame any negative consequences of international trade and investment on the venality of a developing nation's political class. Good governance, to all intents and purposes, amounts to neoliberal governance in the contemporary world. It is thus domestic governmental dysfunction that is blamed for any suffering on the part of the affected communities in developing nations, rather than the exploitative impulse that motivated foreign investment in the first place. When the global condition of the environmental commons is considered in this light it is clear that any credible attempts to undermine the causes of climate change must be thoroughly international, rather than local or regional in nature. This requires, among other things, that the specious logic of environmental national sovereignty be seen for what it is, namely a populist rhetoric designed to mask intrinsically uneven patterns of global development.

In a word, both the causes and the effects of climate change are global in nature. Presently, however, the global environmental movement exhibits a startling heterogeneity with few underlying affinities. This is the case even at the national or regional scale. No doubt part of this variety is due to the fact that historically environmentalism has always meant different things to different people. In this sense, there has never been an environmental 'movement' in the way there was, historically, a relatively coherent socialist movement. Considering environmentalism through the customary lens of left/right politics quickly reveals that there can be conservative or reactionary environmentalism just as readily as liberal or progressive environmentalism. To take a high profile example, the establishment of national parks in the early twentieth century in the United States appeals to liberals as a seminal example of protecting nature from untrammelled capitalist exploitation. At the same time, however, the 1964 Wilderness Act that gave a more extensive policy framework to areas of outstanding environmental value has been cogently criticised for relying on a colonial appreciation of occupied lands as devoid of indigenous populations (see Cronon 1995).

More generally, the debate around environmental protection across liberal democracies is mostly restricted to the confines of weak sustainability. Sustainability, considered as neoliberal environmentalism, holds that natural resources can only be validly measured as material for economic growth. Thus, neoliberal governance is constructivist with respect to nature in a very particular way: it holds that there is no such thing as nature in isolation from economic productivity. This amounts to saying that nature is functionally equivalent to natural resources for which an economic value can be established. It should be recalled, as it is not readily recognised, that the neoliberal paradigm across liberal democracies has long since enveloped both sides of the established political spectrum, left and right.

As a consequence, mainstream environmental political discussion can only differ with regard to the economic calculations but cannot challenge the fundamental neoliberal assumption whereby nature is, more or less, reduced to its contribution to economic output. For example, both sides of the typical western political spectrum can back the development of renewable energy infrastructure, progressives simply being more willing to grant government a role in bringing such infrastructure into existence. What neither side will be able to countenance is the sense that environmental issues call for a significant reconfiguration of both domestic and global economies. Since the time of the publication of the Brundtland Report (United Nations 1987), the SD paradigm has effectively removed all alternative environmentalisms beyond the pale of plausible policy making.

In the first chapter of this book I will explore recent theory that offers resources to overturn the neoliberal appreciation of nature. Broadly speaking, this theory is constructivist with regard to nature, that is, nature is not taken to be something in itself but rather seen as something to which humans stand in relation. This does not amount to denying the reality of nature, but does involve rejecting that strand of modern environmental thinking most clearly exemplified by deep ecology. Deep ecology, first outlined by the Norwegian philosopher Arne Naess (1973), calls for a kind of dissolution of the human ego into nature. The uppermost moral principle is thus to overcome the sense of separateness between the human individual and the natural environment. Indeed, the very idea of nature as environment, that is as the physical surroundings of human life, is seen as a barrier to truly ecological consciousness.

Against this, the theories that I will pursue in the following chapter overlap in the conviction that human life and understanding are made possible by a complex relation to nature. This relational understanding does not lead to deep ecology's call to dissolve one's sense of self in nature. But it does lead, particularly in the work of Latour, to recognition that politics can never be a purely human affair. On the contrary, politics is about how to organise the complex interrelations between humans and non-humans. This in turn involves overcoming the deeply held conviction that only the former can be regarded as political agents. Recognising non-humans as genuine agents, Latour argues, is the truly revolutionary thought towards which modern environmentalism has been reaching for the last four decades. Recasting ecology as the study of the systemic interactions between human and non-human agents means that environmental science can longer enjoy its status as essentially non-political in nature. Quite the reverse: environmental understanding is arguably the most radical form of political understanding there is. Thus, our first task here will be to move towards an understanding of nature as intrinsically and radically political. On this basis we will be able to tackle the question of environmental political praxis as a response to climate change.

CHAPTER I

Political natures

CLIMATE CHANGE AND NEOLIBERALISM

Among progressive environmentalists there is a growing chorus of catastrophism. Ominously, this penchant for labelling climate change a natural catastrophe is becoming equally plausible among the chief agents of neoliberal governance. This is not a confluence opponents of neoliberalism should welcome. Instead, it tends to block out the alternative political vistas potentially opened up by efforts to undo the root causes of climate change. Rendering climate change a natural catastrophe serves to remove it from the social sphere properly speaking and present it as a merely empirical state of affairs, a fact of nature. But climate change is symptomatic of a social-political condition, not an independently existing condition of the natural world.

While it is hard not to get caught up in the urgency to act in the face of climate change, there are many good reasons to resist the condition of panic that is being induced. Chief among these is the readiness with which this urgency of action feeds into the security rhetoric of the neoliberal state. Clearly, the prospect of catastrophic climate change is not bringing about a crisis of the current configuration of the nation state. On the contrary, the scientific and technical complexity of climate change is serving as a pretext to undermine further the democratic accountability of state action. As will be discussed in this first chapter, there are real indications that policy across powerful neoliberal states is converging towards a position that might be labelled a 'war on climate change'. Attention paid to politically organised climate change denial has tended to keep this reality out of the media spotlight.

If the war on terror was in part a struggle for access to oil, a much larger front will need to be opened up to fight for all the goods needed to maintain western consumer lifestyles given the context of accelerating climate change. This will crucially involve access to agricultural goods, including irrigation, chemical inputs, and food commodity distribution networks. Multinational corporate agribusiness is already well placed to prosecute this war on behalf of leading neoliberal states. And whereas poorer nations can get by without access to their own fossil fuel resources, we have yet to engineer a human being that can survive without food.

Thus, as paradoxical as it may seem at first glance, the temptation to announce a climate crisis should be resisted. Once gripped in the psychosocial atmosphere of imminent environmental collapse, opposition to neoliberalism is likely to lose further ground to the autocratic power of the corporate-led state. Again, those who stand amazed at the apparent inaction of nation states confronted with climate change have failed to grasp the nature and limits of neoliberal politics. Contrary to the views of many progressive environmentalists, the neoliberal state is not helpless and largely passive in the face of climate change. Instead, it has been incessantly active in framing this problem as something only neoliberal politics can adequately respond to.

As the effects of climate instability get more severe, efforts to ensure the smooth operation and expansion of neoliberal development are likely to frame the liberal rights regime as something that needs to be curtailed or even sacrificed to ensure access to environmental resources. This would follow the pattern established by the war on terror. Environmental security for wealthy nations and corporations will increasingly be bought at the price of exacerbated ecological insecurity for poorer states and regions. This invidious trade-off will also be on view within affluent societies, where the neoliberal imaginary is most conspicuous and forceful. Neoliberalism calls on individuals, rather than society via the state, to take responsibility for the risks and opportunities of personal flourishing. Hence the spectacle of post-Katrina New Orleans is indicative for the future of neoliberal society, where the losers at the capitalist roulette wheel pay the ultimate cost for relying on a state that no longer looks upon their welfare as its business. One irony of presenting climate change as a natural catastrophe is that it precisely emboldens a neoliberal security state in an age where state action is meant to be limited to blocking impediments to free market operation. Where the environmental activist sees ecological tragedy, the neoliberal economic agent enjoys the prospect of expanded global opportunity.

In *The Politics of Climate Change* sociologist Anthony Giddens (2011) identifies a key paradox. 'Giddens's paradox' states that everyday actions are unlikely to change until the effects of climate change become so intuitively obvious that there will be little we can do to avert the worst consequences to human life. At first glance, the paradox seems to turn on the level of certainty the average actor can possess about the reality of climate change. But, as Giddens makes clear, it is actually as much if not more about the nature of climate change itself. Because the existence of climate change is a scientific construction based on masses of data collected from a vast number of sites over a long period of time and analysed by myriad international scientific collaborations, it is not available as an object of scrutiny as individual perceptual objects are.

However, while climate change shares the property of being scientifically constructed with many other acknowledged realities, this is not the real source of its paradoxical status. Giddens's insight in truth relates to the lack of political action to stave off the present and predicted future effects of climate change. The putative reason for this absence of political intervention is the lack of intuitive availability of the object in question. But is this plausible? Political action is clearly motivated by any number of abstract notions, social justice being an obvious case in point. What Giddens presents as an epistemic paradox is in truth the symptom of a certain overarching manner of governance, namely neoliberalism. Neoliberalism constructs nature and the natural in a distinctive way, namely in terms of individualised personal self-realisation. Self-realisation within the framework of neoliberalism comes down to each person grasping herself as an open-ended array of economic potentialities. According to the neoliberal perspective, collaboration and cooperation are best seen through the lens of mutual competitiveness. The sphere of competition, the market, is the field in which competitive self-realisation is pursued. The market harmonises the competitive actions of individuals, while leaving future developments and acts intrinsically open.

When this (for now loosely described) context of neoliberalism is used to frame climate change, then Giddens's paradox starts to look quite different. Concerted international and individual environmental action is not absent due to the non-intuitive nature of climate change. Instead, it is a question of what neoliberal governance can admit in advance as a legitimate type of action. Jean-François Lyotard, the French thinker best known for his work on postmodernism, identified this phenomenon decades ago in his book *The Differend* (Lyotard 1988 [1983]). Taking up Ludwig Wittgenstein's idea of the language game – the idea that everyday practice is made up of largely implicit rules that define a certain number of socially legitimate moves – Lyotard argued that there are radically different modes (games) of political representation, such that one mode simply cannot be translated into the other.

Neoliberal governance is a paradigm of the political that was developed expressly to overthrow notions of the planned economy. It acknowledges as valid only those instances of state action in the economy that relate to the legal and security domains. State management of the economy as such is the central proscription of neoliberal governance. This explains why, when it comes to robust policy-making, environmental challenges are predominantly, if not exclusively, seen as issues of security. The stale conceptions of Garrett Hardin's notion of the tragedy of the commons, according to which commonly held resources are destined to be mismanaged by individual players, have long since become the unchallengeable presumptions of neoliberal thinking. Thus, it is the neoliberal focus on the nation state's role in ensuring the rule of law that brings about the restriction of governmental action to resource security. This concern often gets presented in the populist terms of energy independence and as such acts as specious compensation to a public distressed by the pressures of relentless economic globalisation.

This account of where things stand is seldom found in mainstream media. Instead, the preferred variety of explanation is one of a political class viciously separated from true public interest. This diagnosis has a few familiar variants: corporate money warping public policy; politicians lining their own pockets or favouring their own special interests over the general good; and so forth. These causes are used to account for the failings of representative democracy as well as popular disaffection with democratic politics. None is this is unfounded. But the problem is that none of these explanations offers any handle on the specificity of neoliberal governance as such. Despite decades of international environmental agreements and frameworks under the auspices of the United Nations, the richest nation states competitively collaborate to safeguard privileged access to carbon-based energy sources well into the future through military means. This is not a symptom of moral failure on the part of individual negotiators or the political class more generally. It is rather clear evidence of the intrinsic nature and limits of neoliberal governance.

According to such governance it is not for politicians but rather for markets to determine how natural resources are to be exploited and distributed. The role of government is to legitimate the rules of efficient environmental commerce and sanction nation states that refuse to play according to those rules. Clearly, even nation states that have 'renationalised' natural resources (for example, natural gas in the Russian Federation or oil in Venezuela) have done so in order to gain more clout in the context of international trade (see Colgan 2014). In this context, what may superficially look like features of a command economy are in reality quite compatible with neoliberal geopolitics.

The mantra of disaffected electorates across wealthy liberal democracies is complemented by the vain hope that getting good people back into politics could be the answer. But this ignores the fact that neoliberalism presides over political decision-making in a way that is thoroughly impersonal and systemic. Using a moral vocabulary to capture the mechanics of neoliberal governance obfuscates its essential nature and so blocks viable routes to an alternative. It is vital to appreciate that neoliberalism cannot be reduced to a mindset that is a product of certain personality traits. Instead, neoliberal governance sets out rules of legitimate action beyond – or before – any particular individual, party, government, or even polity. To return to the game analogy, politicians under neoliberal governance can no more get to grips with the underlying causes of climate change than a chess player can move a rook diagonally. Even if such a move were made, the dominant players of the game would refuse to recognise it as legitimate.

As I write, the treatment of the Greek government's efforts to move beyond austerity economics might serve as a case in point. The problem, in the terms articulated here, is that the Greek government, in seeking to protect its citizens before it serves international finance, is attempting a move that is ruled out in principle. Under neoliberalism it is not the state but rather the market that is the source of legitimate social innovation. The function of the nation state is limited to providing a thoroughly predictable legal framework within which the bounds of legitimate economic moves can be readily identified and transgressions punished. Thus, the role of the state is intrinsically negative and disciplinary, rather than positive and experimental. Neoliberalism is above all a regime of political self-regulation and self-discipline. Politicians and political parties, for largely rhetorical reasons, may readily resort to the language of novelty and renewal (we have heard a lot about 'change' over the last decade), but genuine political innovation is the last thing one can expect of the nation state in the era of neoliberal governance. This is one reason that has led some theorists to refer to the period of globalisation we are currently living through as 'post-history' (see Sloterdijk 2013 [2005]).

Neoliberalism is a term whose origins go back to a specific school of political and economic theory first developed in Weimar-period Germany in the 1920s. This theory was the guiding light in the post-war West German economy from the 1940s on and migrated to the United States principally through the work of Friedrich von Hayek, before being popularised by Milton Friedman during the Reagan administration in the United States. Foucault's (2008) careful reconstruction of the early intellectual history of German neoliberalism (known as 'Ordo-liberalism' from the journal *Ordo* in which many of the seminal articles were originally published) makes clear how neoliberalism sought to stake out a form of governance that opposed both the centralised command economy espoused by state socialism and the potential social chaos engendered by unregulated markets. Government was seen as the vital instrument to ensure stable capitalist economic growth and all governmental action was to be pursued exclusively for such an end.

The recurring motif of Foucault's analysis of neoliberalism is the question of how not to govern too much. In this regard, the neoliberal conception of governance generates a kind of paradox. This paradox takes the form of intervening in the capitalist economy in order to ensure that such intervention is minimised. Management of wilderness might be an apt parallel here, insofar as this involves employing technological (non-natural) means to allow habitat to remain sufficiently natural. Similarly, the goal of neoliberal governance is to facilitate and maintain capitalist markets in their 'natural' form. In this way, as Foucault (2008) effectively demonstrates, through the lens of neoliberalism the market is both naturalised and denaturalised at the same time. It is naturalised as the fundamental context of human action and deliberation, and yet subject to the highly complex, 'unnatural' management of economic policy instruments. As paradoxical as this constitution of the market may seem, this has not prevented the neoliberal political vision from becoming unquestioned common sense for most economists, politicians, and voters across contemporary liberal democracies.

THE LIMITS TO LOCALISM

While it is relatively easy to point out the limitations of legitimate environmental politics under neoliberal governance, the task of conceiving a more credible alternative for framing the social challenge of climate change is much harder. As noted, one pervasive appeal in this regard is to localism or regionalism.

According to this perspective, collective environmental action only becomes plausible at the local level, preferably through face-to-face interactions, which are most conducive to the relationships of trust needed for authentic collaboration. At a larger scale this can be presented in the form of regionalism. This tendency is evident in urban planning, particularly in the Unites States, where the idea of planning such things as transportation networks and population distribution beyond the metropolitan scale possesses obvious logistical advantages.

There is also a sense that nation states are not the 'imagined communities' (Anderson 2006 [1983]) they once were and that regional identification is a more plausible contender. For example, the Pacific Northwest, where I have lived for a number of years, is popularly presented as Cascadia: a bioregion stretching from northern California to British Columbia following the Cascade Mountain range. People who reside in this region intuitively appeal to a common sense of place that is missing when trying to identify with the United States as a whole. Environmental localism can also provide itself with a naturalistic basis from a regionalist perspective, insofar as the topography, native species and climate offer a certain uniformity over a defined area.

As David Harvey (1996) has made clear, however, localism and regionalism can just as easily give rise to regressive as progressive political action. Often environmental regionalism implicitly appeals to an exclusionary tribalism that tacitly supports the kind of natural resource anxieties projected by the neoliberal state. For example, the ample water resources of the Pacific Northwest are to be fervently protected from more arid neighbouring regions, such as southern California, that have no legitimate claim to them. At the same time, it is not hard to see regionalism as simply another variant of neoliberalism's distrust of overreach on the part of central government. In fact, the very idea of geographically determined autonomy that is central to the popular appeal of regionalism or localism arguably runs counter to most environmental challenges. While it may be true that a sense of connectedness to one's place of residence can be a powerful source of political will to protect it, it is no less true that environmental localism runs up against insuperable limits in the face of climate change.

For example, if one looks at environmental protections across different states in the United States today it can be readily argued that states such as California act to raise the bar for other states. But the costs borne by industry thanks to tighter environmental regulation can also act as incentives for those businesses to move to other states where such regulation is absent. The contested and largely ineffective status of the Environmental Protection Agency (EPA) is testament to both the need for, and challenges to, federal environmental policy in the United States. Environmental regionalism is thus a phenomenon that belongs to what Neil Smith (2008 [1984]) referred to as the 'uneven development' that is an inherent rather than contingent feature of capitalist development. Applying the thesis of uneven development to climate change generates the uncomfortable conclusion that reaching international environmental policy consensus on the terms of neoliberal governance is simply impossible. Just as air and ocean currents flow across the surface of the globe in virtue of differences in levels of pressure, capitalist development can only exist thanks to patterns of stark economic and environmental inequality.

Returning to Giddens's diagnosis of the central political problem of climate change, we can readily see that it is not really a matter of the average citizen failing to see the phenomenon in question. Instead, it points to the fact that modes of political agency that would veritably rise to the social challenges entailed by climate change will be deemed illegitimate in the terms of hegemonic neoliberal governance. For example, the recent history of radical environmentalism ('eco-terrorism') across liberal democracies makes clear just how much tolerance will be extended to environmental direct action that amounts to more than symbolic protest. Of course, it can always be argued that radical environmentalism needs to be transformed into the more acceptable forms of mainstream environmentalism in order to attain democratic legitimacy. But this simply returns us to the paradox with which we began.

If the effects of climate change by the end of this century will be as dramatic at the IPCC reports predict (see IPCC 2014), any environmentalism willing to operate within the terms set by neoliberal governance is destined to be overtaken by events

before too long. While neoliberalism preaches economic innovation and free trade, the same capacities are explicitly denied
within the political sphere. This is true of political action both in
its bureaucratic manifestation within the nation state and also
in the looser formations of civil society. In fact, the former is
readily harnessed to combat the latter if environmental activism veers into militancy of any sort. Hence, it is wrong to arrive
at the conclusion that local environmental action can achieve
what the nation state cannot. Localism, it may be somewhat
harshly but nonetheless accurately said, is the current opium
of the environmentally conscious masses across wealthy liberal
democracies.

The irony of championing environmental localism – and a
key reason why it receives widespread acceptance – is that it
echoes the neoliberal critique of the state. Localism appears
more authentic than state action thanks to its ethically superior sense of place. While local environmental activism can
undoubtedly bring about salutary environmental change to
a limited degree, at the same time it serves to underscore the
neoliberal conviction that there is little the nation state can
or should do as an environmental agent. Part of my argument
in this book will be that, of all environmental issues, climate
change makes clear the strict limits to environmental localism.
Climate is the ultimate environmental commons and as such
requires thoroughly international modes of political action.
Conceiving such agency in the absence of or in opposition to
the nation state rests on the tacit assumption that neoliberalism
is the sole political paradigm the nation state can conform to.
History would suggest otherwise.

THE STATE AS ENVIRONMENTAL ACTOR

Another aspect of climate change highlighted by Giddens is its
catastrophic nature. This makes political inaction both more
puzzling and more understandable. Environmental activists
tend to see the inaction as unfathomable because the stakes
are so high. Even if someone is relatively unmoved by the prospect of habitat destruction, non-human extinction, and the

resulting diminution in biodiversity, surely the threat to human well-being strikes a chord? At the same time, the very scale of climate change is overwhelming. Giddens acknowledges this latter reaction, but obviously expects policymakers to rise above it. In fact, the force of Giddens's paradox applies primarily to the average citizen. There is a largely implicit assumption throughout his book that experts and politicians can act more rationally than the average citizen when confronting climate change. This naive assumption of relatively unbiased action on the part of politicians and scientists betrays Giddens's failure to get to grips with the realities of neoliberal governance. He also assumes paternalistic benevolence on the part of the political class, whereas the realities of the market economy lay out a horizon of inexorable economic expansion with few concessions to the quality of life of workers.

Clearly, Giddens is broadly applying his same model of 'Third Way' politics that was so influential in the context of the New Labour movement in Britain in the 1990s (see Giddens 1998). This model insists that the differences of party politics can and should be overcome to arrive at a consensus that furthers the interests of all concerned. This approach to politics is opposed to radicalism of all kinds, although it happily speaks the more reassuring language of policy innovation. Giddens (2011) appeals to the need for more long-term planning to address the multiple challenges of climate change, but again relies on the good sense of policy makers and business leaders to achieve this. Of course, he is cognisant of the fact that a credible response to climate change requires international environmental policy agreement. Thus, what might work for domestic British politics in terms of the New Labour project will be undoubtedly more difficult to achieve at the international level. Conversely, the vicissitudes of public opinion make Giddens suspicious of appeals to robust participatory democracy as the preferred form of environmental agency. Giddens is aware that this makes his position run foul of the political vision of most green parties and groups. The appeal to participatory democracy is usually paired with a strong preference for localism or regionalism. What is interesting about Giddens's position is that, while he recognises the need to engage the electorate on the issue of climate change at the personal and intuitive level, he resists at every turn any version of politics that is credibly citizen-led.

Despite these important reservations, I believe Giddens is right to argue that

> the industrial nations must take the lead in addressing climate change and . . . the chances of success will depend a great deal upon government and the state [. . .] for better or worse, the state retains many of the powers that have to be invoked if a serious impact on global warming is to be made. (Giddens 2011: 138)

This position is opposed to that endorsed by those environmental activist groups across liberal democracies who see the solutions to climate change coming largely from local community autonomy. As Giddens demonstrates, even where local autonomy on such things as energy production has been to a large degree realised (Germany is the outstanding example), the nation state as a whole remains very dependent on fossil fuels (lignite coal in the case of Germany). Giddens is also right to highlight the realities of globalisation, where the environmental costs of goods imported to the richest nations do not generally feature in their domestic emissions figures. Energy-independent towns in Germany may be an endearing poster child for many environmental activists, but once again the limits of environmental localism under global neoliberalism are readily revealed.

Giddens's insistence on the need for strong national environmental policy-making is justified. But he significantly underplays the need for liberal democratic governance to change if it is to become an effective agent in promoting environmental policy with enough clout to meet the challenges of climate change. In his brief case studies of the leading economies, Giddens outlines the well-known rifts in American party politics in relation to this issue. Whereas the major parties in the United Kingdom, by comparison, readily recognise the scientific consensus on climate change, politicians and the public in the United States are riven by dissent. Climate change has become yet another item in the so-called culture wars, on a par with the evolution versus creationism schism. Naomi Klein's (2014) book underscores how well funded and influential the climate change denial lobby is in the US. She recognises that the neoliberal animus against governmental interference in markets is the primary cause of the failure to tackle climate change through state action. The political and social changes required to combat anthropogenic climate change would, so the deniers believe, allow the federal government the pretext to push America towards the sort of paternalistic state that runs roughshod over cherished individual freedoms. The irony for Klein is that the deniers have grasped the full political consequences of facing up to climate change much better than many who accept the reality of climate change:

> I think these hard-core ideologues understand the real significance of climate change better than most of the 'warmists' in the political center [. . .] The deniers get plenty of the details wrong . . . but when it comes to the scope and depth of change required to avert catastrophe, they are right on the money. (Klein 2014: 43–4)

In other words, there is no room for comfortable neutrality and gradual reformism when it comes to climate change.

The question that is largely evaded by both Giddens and Klein with regard to credible responses to climate change is how to move beyond neoliberalism. Giddens offers the prospect of simple good sense on the part of liberal democratic policy makers. Klein sees a popular progressive uprising against the corporate stranglehold over national politics. Klein's analysis suffers from her endorsement of the catastrophe narrative. This is a variant on the well-worn Marxist theme of the crises of capitalism, which leads to a sort of political millenarianism, where a sudden revolutionary uprising emerges to install a just world on the ruins of the capitalist edifice. Unlike Marx, however, Klein offers no credible account of how class solidarity can create a political agency powerful enough to transition neoliberal capitalism to something intrinsically ecological in nature.

Klein's perspective on climate change is obviously highly influenced by the thinking of 350.org cofounder, Bill McKibben. Almost three decades ago McKibben (2006 [1989]) published *The End of Nature*, which articulates many of the core arguments of the environmental catastrophism adopted by Klein. Oddly, Klein has failed to make some obvious connections to her earlier work, *The Shock Doctrine* (Klein 2007). The central argument of that book was that neoliberalism, from its origins in South America in the 1970s on, has basically advanced by inducing or taking advantage of economic crisis. In other words, the very phenomenon that orthodox Marxists view as heralding the collapse of capitalism – periodic economic crisis – is actually the source of neoliberalism's relentless expansion and consolidation. The aftermath of the most resent Great Recession is an obvious case in point. Klein's book is subtitled 'Capitalism vs. The Climate'. This disjunction immediately strikes the wrong chord, as the central task here is to understand how the climate, and nature more generally, are understood through the capitalist or neoliberal lens. Just as there is no nature neutrally available to us, so too there is no climate such that it could taken as something in itself beyond any social or political construction.

McKibben's thesis of the end of nature had already grasped this, positing as it does that anthropogenic climate change disallows all talk of an independently existing nature. The old concept of nature as something acting autonomously, as a sort of frame for human existence on the face of the planet, no longer holds. At best, as McKibben eloquently expresses, this older type of nature can be an object of mourning and nostalgia. For my own purposes I would like to adapt McKibben's thesis of the end of nature to the analysis pursued here. While McKibben speaks in a fairly undifferentiated way about human action on the natural environment, I would like to sharpen the claim in the following way: neoliberal governance constitutes nature in specific ways. It is not enough to identify industrialisation as the historic cause of climate change, as McKibben does; nor is it enough, as Klein does, to point out the various regional and local forms of opposition to capitalist exploitation of natural resources. More basic than either of these analyses, it is necessary to grasp the very specific ways in which nature can appear within the paradigm of neoliberalism.

If we choose to follow the catastrophe narrative of climate change, as Klein and McKibben do, then the need to get to grips with how neoliberalism has reconfigured both nature and politics over the last four decades will be rejected on the grounds there is insufficient time to think this through. The consequent conception of environmental action to tackle climate change will be both reactive and inadequate. It is often pointed out that, when polled, voters in wealthy liberal democracies invariably place economic vitality over environmental health. Environmentalists have consistently argued that this preference is ill-considered and short-sighted and yet it persists. During an economic crisis, this preference hardens and environmental concerns drop further down the ranks.

Thus, Klein's hope that the catastrophic implications of climate change will be the catalyst for effective environmental political action seems unwarranted. More dramatically, as the world approaches environmental limits of various sorts in the current century the more realistic prospect is that international conflicts will increase sharply as the leading economic nations resort to more and more flagrant natural resource hoarding. As Harald Welzer (2012) argues in *Climate Wars*, the long history of resource scarcity is closely tied to historic acts of systematic genocide. Given the present global political dispensation, this outcome strikes me as the most plausible. On the other hand, this is not the result of some inherent tendency of human nature but primarily due to the way nature and natural resources are understood and exploited under neoliberalism.

Some of the alternative conceptions of nature propounded by the environmental movements investigated by Klein may point in the right direction. But the ultimate challenge is not about identifying alternative worldviews (contemporary liberal democracies are well provisioned in this regard), but rather a question of realising effective modes of environmental political action. Such action needs to overcome the neoliberal opposition between grassroots activism and state-led political action. The former will always prove too limited unless it succeeds in wielding state power. The action of the nation state is at once the source of neoliberalism's pervasive strength and its greatest potential weakness. But to exploit this weakness, the particular constitution of state action that limits it to facilitating the expansion of the market must be undermined. In other words, the only credible course for environmental politics is a radical reconfiguration of state power.

It is simply not credible to think coordinated local environmental efforts will ever be enough to avert the worst effects of climate change. For every modest gain made by non-governmental initiatives on this front, the global forces of neoliberalisation will have shifted the balance further towards climate instability. On Giddens's account, by contrast, there is a marked preference for national environmental policy-making over grassroots environmental activism. Giddens is sceptical of appeals to political agency emanating directly from civil society without the mediation of government bureaucracy. While the vision of direct democracy gives many progressives a warm feeling, it is arguably inadequate in the face of the complexity of political action in contemporary post-industrial states.

In the next section of this chapter I will consider a further aspect of this complexity when confronting Latour's call for a politics that includes a vast network of non-human agents. But before such complications are introduced, it is important to underscore Giddens's argument for the pivotal importance of nation states in tackling climate change. Despite over two centuries of political theory calling for transnational political rule (stemming from Enlightenment-period cosmopolitanism), we still live in a world where nation states remain the exclusive source of ultimate political legitimacy. The peculiar hope that constitutes the legacy of modern liberal democracies is that representative government can truly act to promote the good of all citizens. Given that climate change threatens to undermine the very material basis on which human flourishing generally rests, it is clearly a failure if the nation state does not do everything in its power to mitigate its effects. The question thus becomes how neoliberal governance permits, or even ensures, this failure. While Giddens is correct to see the nation state as key in efforts to tackle climate change, he drastically underestimates the changes in the constitution of the liberal democratic state needed to confront the problem of climate change.

THE NATURE OF POLITICS

In the introduction to his book *Politics of Nature*, Latour (2004) spells out how the radical political implications of the environmental movement have yet to be appreciated: 'My hypothesis is that the ecology movements have sought to position themselves on the political chessboard without redrawing its squares, without redefining the rules of the game, without redesigning its pawns' (Latour 2004: 5). The reference to game rules and the role of pawns in the game of chess more specifically is telling. The pawn is that element of the game that is seemingly the weakest, most readily removed piece, but also that agent most capable of radical change (that is, into a queen). There are also many more pawns than other types of pieces on the board. This serves as a metaphor for the status of the environmental movement in contemporary politics: the number and diversity of environmental advocacy and activist groups in liberal democracies is bewildering, their concerns more heterogeneous still. Equally, the standing of such groups within mainstream politics has never been settled. When green parties are formed environmental concerns can be more readily absorbed into the bureaucratic decision-making of orthodox politics. But such developments invariably involve compromises that dilute the peculiarity of ecological politics. This is obviously a matter of concern to Latour, who considers the underlying task of political ecology to involve a radical recasting of politics as it is normally understood. My analysis in this book agrees with this important contention. Just as Latour claims environmentalism cannot be woven into the existing fabric of politics, so too I argue that climate change constitutes a social-political challenge that cannot be met on the terms of neoliberal governance.

It a profound irony that the underlying imperative of neoliberal capitalism, namely growth, is a biological metaphor seldom recognised as such. The indicators of economic growth – whether gross domestic product calculated by economists or the share indexes seen and heard daily via mainstream media – mean little to anyone lacking specialised knowledge. On a more basic level, the term 'growth' immediately connotes health and flourishing. A little reflection on the matter, however, quickly reveals that for actual living organisms healthy growth is always limited. Unlimited growth here could only mean that the type of organism or the system within which it grows is somehow unbalanced. The phenomenon of 'eutrophication' (denoting excessive plant growth due to overconcentration of fertiliser in aquatic systems), primarily due to agricultural run-off, is an apt concrete equivalent of the inherent trend of neoliberal capitalism. Beyond the fundamental metrics of economic growth from a neoliberal perspective, there are a number of qualitative features: urbanisation, formal educational attainment, openness of markets to transnational capital, and so forth. It follows that any country where a sizeable proportion of the population is residing and working in the countryside, not availing of formal education, and relying heavily on its own economic production is in need of growth or development.

Another irony here is that, for the most part, the lower the level of development the lower the per capita impact on the environment. As is often pointed out, were the entire human population to attain the level of economic development currently enjoyed by the leading economies, efforts to mitigate climate change would be definitively thwarted. Neoliberal economic development is necessarily uneven, so there is no need to worry that such equality will ever be attained in reality. Equal distribution of wealth is patently not a necessary element of growth considered globally. Thus the recent focus on the extreme concentration of wealth within developed economies is not particularly significant from an ecological perspective (see Piketty 2014 [2013]). From a political ecology perspective, the key point is not to spread the rewards of neoliberal growth more equitably, but to overturn the very concept of growth that is in play. This, in turn, requires a return to the fundamental socialist critique of minority ownership of production under all forms of capitalism.

Returning to Latour, his own particular version of rewriting the rules of politics has to do with overcoming the constitution of the human being as subject. Indicated most clearly by Aristotle's definition of the human being as the political animal, this ensures that politics is thought of as an exclusively human affair. In face of this restriction, preserved through the early modern scientific revolution and Enlightenment constitution of the social, the challenge of environmentalism is the inclusion of non-humans into politics. The constitution of modern politics according to Latour is based on a fundamental bicameralism according to which society and nature are rigorously separated. Because environmentalism seeks to bring non-human entities ('nature' in one sense) into the sphere of politics it challenges this fundamental separation.

The difficulties of maintaining the division are also apparent in efforts to tackle environmental issues through the means available to normal politics. Typically, struggles over natural resources are dealt with by identifying and then weighting the interests of various human stakeholders. Within these calculations non-human entities are considered things onto which human value (in the form of 'interests') are projected. This strongly taints non-human entities with a basic sense of unreality. More fundamentally, it reduces ecological politics to the ambiguous status of various human perspectives, all valid in themselves and so difficult if not impossible to compare. The usual result, as environmental activists know all too well, is that the environmental perspective finds itself superficially acknowledged but practically ignored. The subordination of ecological concerns within neoliberal politics has consistently tempted environmental advocates into complicity with the modern nature/society dichotomy. Accordingly, ecological politics seeks to serve 'nature', restore natural systems, or even promulgate rights for Mother Nature. Confronted by a politics that refuses to countenance the very idea of non-human agency, environmentalism attaches itself exclusively to its side of the divide: nature divorced from politics.

Against this tendency, Latour insists that *'political ecology has nothing to do with nature*. To put it more strongly, at no time in its short history has political ecology ever had anything to do with nature, with its defense or protection' (Latour 2004: 5). Latour qualifies this startling claim by adding that what environmentalism says it is doing (defending nature) is in contradiction with what it actually does, namely attempting to bring non-human things into the sphere of politics. It is important to note that Latour has a very expansive conception of politics as '*the entire set of tasks* that allow the progressive composition of a common world' (Latour 2004: 53). The nature/society division that is constitutive of politics in the orthodox sense brings with it a further separation between science and politics as domains of knowledge. While nature can only be rigorously known by science (and so not by the average person lacking expert knowledge), politics is reduced to rhetoric and interest, that is, to a play of subjective sentiments with little to no purchase on material reality.

The relevance of this analysis can be readily recognised in the protracted political debates over climate change. On the one hand, there are the environmental campaigners who impugn politicians for ignoring 'the science' and simply pandering to special interests or populist sentiment; on the other, there are the climate scientists, who insist they will have nothing to do with any 'politicising' of their scientific research. Any number of other disciplines step forward as potential mediators for this stand-off: most notably psychology and cognitive science, claiming to reveal how inaction with respect to climate change is neurologically predetermined; ethical theorists, mostly pursuing the line of duties to others (including future generations) neglected by allowing climate change to go unchecked; or economists, who insist that rational choice will always prefer short-term individual satisfaction to long-term collective management.

Latour's more general point is that environmentalism points towards a definitive overcoming of the nature/society, science/politics, and human/non-human divisions. In terms of neoliberal governance, his argument can be understood as questioning the coherence of the sustainable development model. According to this model the underlying matter of concern is meeting the basic needs of the global human population. Against this, environmentalism has all along asked: what are human needs taken in isolation from the ecological systems in which those needs arise? Latour complements and echoes this by asking: what is human politics taken in isolation from non-human things?

This leads Latour to the question of political representation. He alludes here to the political principle of universal enfranchisement according to which those affected by political decisions should each have their own say. Political ecology, Latour argues, basically raises the question of how to allow self-representation for those non-human entities with which environmentalism is concerned. He is aware, of course, that the time-honoured conception of politics only admits human speakers into the political chamber. Accordingly, speech attributed to non-humans is at best possible by raising mere objects to the status of honorary subjects, possessing capacities and rights only by analogy. This kind of politicising of the natural is familiar within contemporary environmentalism but it is not what Latour has in mind. Instead, he insists that ecological politics calls for a form of democratic human/non-human organisation that is non-representational in nature.

The natural sciences, for Latour, are the primary means by which non-human things achieve representation. But again the conventional view is that these sciences establish only the facts about reality. Politics, by contrast, is the play of interests between human subjects and their relatively organised factions. Again, on Latour's interpretation, political ecology is not striving to be a special interest – comparable to corporate lobbyists – but rather calling for a complete overhaul of the political system, a revolutionary constitutional change. He sums up the problem: 'How are we to go about *getting those in whose name we speak to speak for themselves?* By refusing to collaborate, political philosophy and the philosophy of the sciences has deprived us of any opportunity to understand this question' (Latour 2004: 70). Latour is clearly arguing that the political implications of the modern environmental movement have been vastly underestimated. Environmental politics is not normal politics that has integrated a concern for nature. Instead it is politics reformulated from the base up.

The principal significance of Latour's analysis to my argument in this book is this: it makes clear that no amount of reform possible within the schema of neoliberal governance (as the latest variant of capitalist modernity) will be sufficient to meet the social challenges raised by environmental concerns. This holds *a fortiori* in relation to climate change, given it is generally recognised as the greatest current ecological challenge. This contention is a variation on my earlier claim that Giddens fails to recognise the extent to which the politics of nation states would have to change to confront climate change in credible ways. Presently, nation states enter climate negotiations on the terms established for trade agreements. It is a given that each state will pursue exclusively its own interests and will cooperate with others only to the degree that these interests are served or at least not obviously hurt. Political good sense dictates that the only sound environmental policy is an instrument that promotes rather than damages economic growth.

The concept of sustainable development employed in this context grasps non-human things only through the lens of human needs, as if these needs were somehow solely attributes of human subjects having no reality beyond the subject. This translation of non-human things into human needs (that is, as 'natural resources') is further subject to the ruling market schema, according to which a need is defined by any good supplied by the market. In this context any division between basic and non-basic needs becomes highly relative and so contestable, thereby disallowing the kind of moralising distributive arguments put forward by environmental ethics or standard arguments of redistributive social justice. Against all this I believe Latour's conception of political ecology as the effort to attain non-human self-representation is a fruitful one to pursue. It makes clear how radical political reform is needed to tackle climate change, thereby offering a way forward beyond the familiar impasse of local versus global action, grassroots versus bureaucratic democracy.

I have introduced Latour's analysis of political ecology here not because it offers any explicit critique of neoliberal governance. That critique can be found in many other thinkers considered in this book, thinkers who mostly follow the critical Marxist tradition in one way or another. For his part Latour regards Marxist dialectics and such notions as commodity fetishism as falling fairly and squarely within the modernist paradigm and its fundamental subject/object dualism. Many efforts have been made to construct an environmental politics on Marxist terms. In the next section one such effort will be explored when we turn to Neil Smith's work on uneven development. Latour's analysis of environmentalism is very far from offering a concrete programme of environmental action. It does not use the vocabulary of environmental justice according to the distributive paradigm. It does refer to political representation, but not in the familiar guise of the Global North exploiting the natural resources of the Global South without any democratic mandate.

More basic than this, Latour carries out an ontological inquiry that asks whether nature and society are really two intrinsically different things. He starts from the observation that, in everyday media representation, the facts established by the natural sciences inevitably get caught up in political and social controversy. For Latour this indicates that science can only be regarded as value-free thanks to a traditional prejudice that has become more challengeable the more environmental concerns have moved to the centre of public debate. Whether the issue of controversy is GMOs (genetically modified organisms) in agriculture, geo-engineering, or the atmospheric effects of deforestation, environmentalism is a scene of politics where facts and values are inextricably linked. Latour's point is that, increasingly, all science is being drawn into politics, indicating that science is intrinsically political in nature. This conclusion holds pre-eminently where economics passes for a science in a strict sense. Here the older term for the discipline, 'political economy', attains a renewed relevance.

THE PRODUCTION OF NATURE

Neil Smith originally published his book *Uneven Development: Nature, Capital, and the Production of Space* in 1984, with substantial new material added for the third edition from 2008. Smith's starting point is the notion of the production of space as developed by the French theorist Henri Lefebvre (see Lefebvre 1991 [1973]). Given the crucial social impact of urban development within the capitalist economy Lefebvre sought to place such development at the centre of Marxist political analysis. This involves challenging the more orthodox focus on time and history in favour of a complementary critical spatial analysis. In *The Limits to Capital* (2006 [1982]), Smith's intellectual mentor, David Harvey, had exhaustively applied Marx's economic analysis to the problem of fixed capital and rent extraction. Similarly, *Uneven Development* starts out with the recognition that the 'spatial immobilization of productive capital is no more or less a necessity than the perpetual circulation of capital as value' (Smith 2008: 6).

As Harvey has consistently pointed out, fixed capital represents a fundamental problem for commodity capitalism, insofar as there are limits to making real estate fully subject to the accustomed cycles of obsolescence and replacement. The concrete solutions to this dilemma are to be found primarily in the manifold applications of urban renewal. Such renewal, of course, runs the gamut from the preservation of historic buildings, to modernisation or retrofitting, to wholesale demolition and replacement. The means employed in urban development are equally various and yet the underlying goal is one and the same: the intensified extraction of ground rent.

Neoliberalism, considered as the dominant political paradigm of the last forty years or so, has been crucially characterised by a global reinvestment in urban centres. The causes of decay within the central city prior to revitalisation are extremely various, involving deep economic, demographic and political factors that play out very differently from place to place. The solutions sought within urban planning have also been diverse. And yet it has emerged ever more clearly in recent decades that urban redevelopment is a crucial element of the neoliberal economy. This accounts, at least in part, for the popularity of localism discussed earlier with regard to environmental activism. Cities are intensively marketed to actual and would-be residents alike in a variety of direct and indirect forms.

This points to the basic fact that cities are locked into relationships of mutual competition over economic resources such as business investment, skilled labour, and so forth. The apparent attractiveness of the urban environment is an obvious factor within this competitive framework. The centrality of such development within capitalist economic dynamics is captured by Lefebvre's notion of the production of space. According to a common sense understanding, material production requires a given place or space within which it can occur, but space itself it not thought as something produced. Lefebvre's breakthrough was the recognition that capitalist production must also, crucially, produce the space in which it can take place.

While Marx and Engels were of course cognisant of the fact that modern industrialisation gave rise to the massive urban centres of capitalist production, the focus of their analyses was on the circulation of capital and on capitalism's pre-eminent ability to remove all fixed geographical barriers that served to slow down this circulation. Lefebvre's notion of the production of space acknowledges that capitalist flows are always caught up in a process of relative geographical fixation: as well as fluidity there is a certain fixing in place of capitalist accumulation. This phenomenon can be readily recognised by recalling the real spatial concentration of financial markets within a handful of iconic affluent cities: New York, London, Frankfurt, Tokyo, Hong Kong and Shanghai.

This concrete localisation of international finance, on Lefebvre's analysis, points to the fact that capitalism can only operate by producing real spaces in which to anchor its dynamics. As the process of post-industrialisation took hold in the 1970 and 1980s, the wealthiest, most developed economies were obliged to derive significantly increased levels of value from real estate. In accordance with the production of space conception, this was the real cause of the 'urban renaissance', counter to the more orthodox explanation of a sudden reversal in the sentiments of the affluent middle class with regard to urban life. Now decades into this process, gentrification in certain cities across wealthy liberal democracies has reached crisis proportions, such that the workers needed to keep the urban machine going are forced to live at greater and greater distance from their places of work. Competition over urban resources is a key element of the neoliberal reconfiguration of everyday life. Access to environmental goods such as green spaces, healthy food, as well as educational infrastructure are familiar elements of this struggle within neoliberal urban life.

Smith's (2008) argument is that the production of space is founded on a more fundamental production of nature. Akin to those environmental thinkers who criticise the construction of nature in terms of pristine wilderness, Smith sees resistance to the idea of the production of space as founded on the idea that nature and society are two radically distinct categories. Smith argues that we generally take space as something given and natural rather than socially produced. The central claim of neoliberalism that the capitalist market is a direct result of human nature is grounded in the assumption that capitalist commodities invariably arise to meet genuine human needs. Why else would a commodity exist? Smith's answer is simple and direct: 'Capitalist production (and the appropriation of nature) is accomplished not for the fulfilment of needs in general, but for the fulfilment of one particular need: profit' (Smith 2008: 78). Smith cites Marx's observation that under capitalism, for the first time in human history, nature is appropriated for production to a universal, global degree. In this sense, recognising a global condition in which all material nature is brought into productive cycles, Marx anticipates the death of nature thesis. Under capitalism, nature is definitively 'socialised', that is, can no longer be seen as something existing independently of human social action.

Smith highlights an important distinction between the production of and mastery over nature. The former but not the latter, he observes, is the underlying concern of capitalism. This explains how it can be consistent with capitalist development that a host of undesirable environmental by-products are produced. One cannot say that these are, as such, intentional results of capitalism. But neither are they avoidable. In economic parlance, they are the negative environmental 'externalities' of capitalist production. Much contemporary work within the field of environmental economics comes down to efforts to internalise these externalities in the form of 'full-cost accounting'. The naivety of this approach is all too evident from the perspective of Smith's analysis. Capitalist production will not limit the drive towards increasing profit unless some countervailing social or political power obliges it to do so.

The contemporary neoliberal orthodoxy holds, by contrast, that environmentally sustainable solutions can only be sought among the entrepreneurial class itself as primary agents of market coordination. This approach generally favours further rationalisation of production and intensified consumption patterns. A typical environmental 'solution' that arises here is energy efficiency, which generally requires some sort of centralised administration. But the consumer logic dictates that cheaper prices should be perceived as opportunities for greater rather than less consumption (what economists refer to as the Jevons paradox). Consuming more because something is cheaper is likely to result in little to no net environmental gain. More profoundly, the neoliberal sustainability paradigm leaves unanswered Smith's main contention, namely that the capitalist production of nature is fundamentally uncoupled from the concern for human flourishing (or 'socially necessary' production). Interestingly, in material added to the 2008 edition of his work, Smith explicitly acknowledges climate change as a significant example of capitalism's dysfunctional production of nature (Smith 2008: 80). With the benefit of hindsight, we are able to comprehend much more readily than was possible in the 1980s how the issue of climate change, of all aspects of the capitalist production of nature, is the most problematic. The failure of neoliberalism to innovate its way out of the problem of climate change points to a fundamental limit of this paradigm of governance.

So far we have considered Smith's notion of the production of nature as the foundation of the production of space but said nothing about his articulation of uneven development. Broadly speaking, uneven development is a result of the inherent tendency of capitalist development towards a decreased rate of profit. In an urban context this can be readily observed through the process of gentrification, where initially low values attached to businesses and residential housing stock within a particular neighbourhood dramatically rise until a certain value ceiling is reached, at which point further redevelopment becomes insufficiently profitable to attract additional investment. The focus of investment will then move elsewhere until such time that the real estate values of the original neighbourhood have sufficiently decreased to make reinvestment viable again.

One of the conclusions to draw here is that in this pattern of development the equalisation of value is ruled out. In fact, it is the very unevenness of the development that marks capitalism at the most fundamental level. While this is true at every geographic scale, it is necessary to add that the wealth gaps are relative within, and not absolute across, scales. Recent urban development in wealthy liberal democracies has arguably made the contours of uneven development much more evident. As the distribution of wealth has narrowed the market value of homes has exploded, particularly in places where employment opportunities are more abundant. The same period has borne witness to the casualisation of employment, where both the longevity and conditions of many jobs have become more uncertain.

In recent decades personal debt (including mortgage debt) has also risen exponentially under these conditions. Persistent encouragement to view one's home primarily as an economic investment rather than a means to satisfy a basic human need meant that, when the global housing markets collapsed after 2008, many households were inevitably faced with the prospect of foreclosure and a return to the rental market. The customary false alternatives of moral analysis (greedy bankers or feckless homeowners?) suffused mainstream media as public opinion went in search of a guilty party. The usual platitudes of human greed and lack of foresight were the predictable result of this bogus investigation. The political class chimed in with harsh words for predatory lenders and a banking sector out of control. In reality, the aftermath has been no different from any other scenario of disaster capitalism, other than the sheer speed and scope of the transfer of wealth upward achieved by austerity politics.

CONCLUSION

The core argument offered by this chapter is that neoliberal governance corresponds to a historically particular production of nature. Sustainable development (SD) is an offshoot of this social-political paradigm and so cannot offer solutions that run counter to such governance. Neoliberal capitalism's basic mechanism of uneven development means that wealth is inevitably concentrated rather than distributed across society, considered regionally, nationally or globally. This means, in turn, that the displacement of the negative environmental impacts of production from more to less wealthy geographical locations is an inherent feature of capitalist dynamics. This displacement will always be the preferred 'solution' to environmental externalities under capitalism.

Because climate change involves the most basic ecological commons at a global scale, we can say with certainty that neoliberal capitalism is incapable of meeting the environmental challenges it generates. While this does not rule out exceptional efforts on the part of individual nations, corporations or civil society organisations, the idea that such examples could be generalised is not credible. It needs to be underscored, however, that it is not the modern bureaucratic state as such that blocks the way to real action against climate change, as proponents of environmental localism often contend. There is no reason, in principle, why nation states could not agree to internationally enforced policy that would mitigate and eventually reverse climate change. What fundamentally prevents this is the particular determination and limitation of state agency within the neoliberal paradigm.

Thus, this first chapter has clarified a number of key points that will be expanded upon in what follows:

1. Climate change is an environmental challenge that can only be met when the nation state acts as an effective environmental agent.
2. The failure of the nation state to meet the challenge of climate change is not inevitable but rather the result of its neoliberal configuration.
3. Reconstituting the state to meet the challenge of climate change requires a new politics that moves beyond the modernist division of nature and human society.
4. Climate change is a symptom of neoliberal governance that indicates an uncoupling of the production of nature from the basic social necessity of ecological well-being.
5. The SD paradigm is environmental neoliberalism and ultimately constitutes a dead-end for efforts to mitigate and reverse climate change.

The following chapter will build on these points in order to establish these further underlying contentions:

6. Given the protracted task of transforming politics to meet the challenge of climate change, invoking imminent environmental collapse is counter-productive.
7. The rhetoric of natural catastrophe will inevitably allow the current neoliberal policies pursuing resource security to remain the only viable environmental politics.

It is readily apparent that contemporary environmental portrayal of climate change is increasingly catastrophic in tenor.
Climate change activism, spearheaded by groups such as 350.
org, relentlessly deploys a rhetoric of imminent ecological collapse. While this well-meaning message is consciously designed
to upset perceived complacency by impressing the need for
urgent action, it is apt to misidentify both the nature of the
problem and credible responses to it. Recalling Latour's insistence that nature and politics are always intertwined, it is easy
enough to recognise climate activism's desire to depoliticise the
issue of climate change. At first glance this may seem a strange
judgment to make. Isn't climate activism all about challenging
the political power of the fossil fuel industry and politicians who
place economic interests above ecological health? My point,
however, is that the basis on which the struggle against vested
interests is waged is too narrow a political platform. Let us sup-
pose that the climate activism movement is successful within
the next decade or so. How will the world have changed? The
change will amount to a shift from fossil fuel based energy to
some mix of renewable energy sources. This will do nothing,
in itself, to overturn the basic capitalist imperative for com-
pound growth across the global economy. If fossil fuel based
profit-making dries up other avenues must be found. These will
inevitably involve their own systemic negative environmental
'externalities' as well as the essential damage to workers based
on the familiar patterns of oppression and dispossession. It is in
this sense that the politics of environmental catastrophism are
intrinsically inadequate. The next chapter explores this issue in
detail.

CHAPTER 2

Nature's ends

CLIMATE CHANGE AS NATURAL CATASTROPHE

The previous chapter established the central contention that climate change is a symptom of neoliberal governance. This chapter advances the further claim that the tendency to present climate change as an impending natural catastrophe is likely to blunt efforts to undermine neoliberal governance. Catastrophic portrayal of climate change has become so common that questioning its legitimacy requires a deliberate effort of mind. It is important to be explicit about the fact that my critique of climate change catastrophism has nothing in common with climate change scepticism, an attitude adopted by many on the right. I see no reason to call into question the basic findings of the IPCC and the credibility of the future climate scenarios it predicts. Rather, my immediate concern is with those on the political left who feel that rendering climate change a matter of imminent ecological collapse is the best way to galvanise efforts to mitigate or reverse it. An example of this approach is offered by Naomi Klein (2014) in her recent book *This Changes Everything: Capitalism vs. The Climate*. In the introduction Klein sets out her basic position:

> Climate change has never received the crisis treatment from our leaders, despite the fact that it carries the risk of destroying lives on a vastly greater scale than collapsed banks or collapsed buildings. The cuts to our greenhouse emissions that scientists tell us

are necessary in order to greatly reduce the risk of catastrophe are treated as nothing more than gentle suggestions, actions that can be put off pretty much indefinitely. Clearly, what gets declared a crisis is an expression of power and priorities as much as hard facts. But we need not be spectators in all this: politicians aren't the only ones with the power to declare a crisis. Mass movements of regular people can declare one too. (Klein 2014: 6)

Calling for a conversion of climate change into what she calls 'a people's shock' refers us back to Klein's (2007) earlier book, *The Shock Doctrine*, which exposed the way neoliberal capitalism has established itself over the last four decades by taking advantage of periodic economic and social crises. This reminds us of the origins of the neoliberal reorganisation in the protracted crisis of the world economy from 1973 on.

Taking Klein's stance as typical of a much broader tendency of left-leaning environmental politics, as I see it the problem resides precisely in the catastrophic framing of climate change. To take a complex phenomenon such as climate change and interpret it as a crisis is almost certain to have one basic upshot: the perpetuation and intensification of the hegemonic neoliberal order. And this is what we are already seeing in strategic policy-making across leading neoliberal polities. An example of this from the United States Department of Defense will be explored later in this chapter. Unlike the climate sceptics Klein takes such delight in pillorying, it is clear that the US government is perfectly aligned when it comes to the climate catastrophe hypothesis.

This is a manifest problem for Klein's invocation of a popular crisis. When it comes to organisational readiness, clearly the US military is much further advanced than the kind of civil society environmental groups Klein sees as the vanguard of ecological activism. Needless to say, there is an antagonistic relationship between these two constituencies of the catastrophe agenda: militarised responses to climate change remove the state further away from democratic accountability and justify paternalistic and oppressive environmental politics. Consequently, as I argue in this chapter, progressive environmentalism is inadvertently strengthening the reactionary state security agenda by proclaiming an impending natural catastrophe.

There are, as I see it, three principal reasons why the natural catastrophe framing is not helpful to the political left. First, the Marxist heritage of 'crisis theory' is historically exhausted, particularly in the context of contemporary liberal democracies. Historically, popular leftist attempts to induce or take advantage of a crisis (the Paris Commune, for example) have tended to bring about a more virulent version of the presiding regime. Second, climate change is a symptom of neoliberal capitalism and not its underlying cause. It follows that, even if a popular movement were somehow capable of mitigating or even reversing climate change, all the profound social harm caused by capitalist patterns of production would remain in place. Third, the uneven global development necessarily engendered by neoliberal capitalism allows the main agents of neoliberal governance to shelter themselves more effectively from the worst effects of climate change. This significantly reduces the likelihood that announcing a climate crisis is likely to affect this class of key decision-makers.

The most powerful neoliberal states are likely to use the climate catastrophe hypothesis to justify more blatantly unfair access to vital natural resources. Even within wealthy liberal democracies such patterns of oppression and harm will become more pronounced, as in the case of the social and racial cleansing evident in post-Katrina New Orleans. Klein (2015) herself has highlighted this very thing. While Klein shows obvious relish when presenting the hubris of neoliberal entrepreneurs bent on geo-engineering solutions to climate change, her account of local and indigenous environment activism fails to offer a convincing David versus Goliath story. What fails to emerge in Klein's account of capitalism's confrontation with the environment is how precisely to present climate change as a 'contradiction' of capitalism in a way that truly looks beyond neoliberal governance.

What Klein fails to recognise is a point argued for in the previous chapter: namely that environmental localism knits more or less seamlessly into the neoliberal fabric. It would be nice to believe that, as Klein claims, '[f]ree market ideology has been discredited by decades of deepening inequality and corruption, stripping it of much of its persuasive power' (Klein 2014: 465). But there is little evidence for this in reality. The fact is that, following the Great Recession, wealthy liberal democracies have never been more completely subjected to neoliberal discipline. If awareness of climate change is to alter this basic social reality it will not be on account of the 'fact' that an environmental crisis is underway. Quite the reverse. Rather than predicting an imminent ecological collapse in the course of the present century, it is necessary to see the long social history of capitalism as one long, drawn-out catastrophe. A catastrophe that has become so naturalised within a pervasive common sense that envisaging an alternative social reality has become all but impossible. Climate change is not the fundamental reason to overthrow neoliberal capitalism, but rather the latest symptom of a neoliberal political 'crisis' that has prevailed for decades.

Taking climate change as a symptom of neoliberal governance does not imply that alternative forms of governance guarantee better environmental results. As is commonly known, the former Soviet bloc had a far worse environmental track record than wealthy liberal democracies in the latter half of the twentieth century. That the environment could be handled worse than it currently is under neoliberalism is not in question. What is at issue here is whether environmental neoliberal governance in the form of sustainable development is capable of reversing climate change. My argument is that, due to the intrinsic nature of neoliberal capitalism, it is not. If this incapacity were obvious then presumably the levels of commitment to SD on the part of national governments, powerful corporations, as well as environmental pressure groups and activists, would be all but unaccountable.

The historical peculiarity of neoliberalism is that the propagation of its social imaginary undermines both the anarchist critique of the state apparatus and the socialist demand that the means of production be taken into public control. Despite the rhetorical presentation of its agenda, neoliberal governance does not render the state redundant as a social-political agent. But it does systematically restrict state agency to action that catalyses the self-entrepreneurialism of all members of society. This tendency is to be observed in all domains of contemporary life: in the move away from steady employment to precarious self-employment; from universally available state education to individually chosen and privately managed schools; and from traditional political party affiliations to floating or protest voter status. In each case fragmentation of the public sphere is proclaimed as progress for liberated consumer choice.

The greater array of personal options is combined with a seemingly contradictory tendency towards corporate consolidation. In all productive spheres, whether food production and retailing, manufacturing, medicine, media, or communications technology, the market dominance of a handful of massive transnational corporations grows year on year. In addition, growth in profits is accompanied by greater concentrations in wealth. While the financing of public services is subjected to a harsh regime of austerity, the major corporations are awash with wealth. This abundance and concentration of private wealth is only possible at the expense of the commons, first and foremost the environmental commons. Capitalist efficiency is an imperative that holds only in relation to the extraction of profit. When it comes to the environmental commons, neoliberal capitalism is wasteful in the extreme.

THE END OF HISTORY AND THE ENVIRONMENTAL SUBLIME

Of all contemporary environmental challenges, climate change has emerged as the greatest and most complex. The argument advanced here is that this complexity should be regarded as essentially social-political in character rather than scientific-technical. While science posits environmental limits beyond which we cannot safely go (most commonly, a two-degree Celsius global temperature rise), public debate does not tend to dwell on the political causes of climate change. The neoliberal social imaginary projects a Promethean entrepreneurial optimism that reflexively assures us there is no problem human ingenuity cannot solve. But climate change is not a mathematical puzzle that some corporate or state sponsored team of brilliant scientists can neatly unravel. It is in essence a collective action problem.

The neoliberal perspective allows for this only in the form of consumer behaviour modification. The naive psychology of orthodox economics posits that consumers only ever act on their own, more or less enlightened, self-interest. If climate change is the result of self-interested consumption choices, then it follows that consumers will alter their patterns of choice as the negative effects of climate change become evident. This line of thought has recently given rise to the political mythology of the 'green economy'. At one time heavily touted by Obama as America's way out of the Great Recession, green jobs are said to offer both a way to revive the domestic economy through job creation and to help reverse or mitigate the effects of climate change. Once the global realities of free trade allowed Chinese manufacturing of solar panels to undercut US production, state investment in green jobs could be pilloried by the Republican opposition as yet another example of governmental economic blundering.

The fundamental point, however, is that climate change cannot be tackled through economic growth, 'green' or otherwise. Neoliberal governance has been extremely successful in promoting the growth for growth's sake agenda. It has also engendered a social condition under which the average worker is obliged to operate with a work for work's sake attitude. An irony of the neoliberal order is that, in an age when government has abandoned the task of guaranteeing full employment, the highest political priority of the electorate across liberal democracies is job creation. But neoliberal economics is fundamentally uninterested in generalised, equitable accessibility to well-paid jobs. The social damage of neoliberalism has been amply demonstrated in liberal democracies over the last forty years through patterns of systemic unemployment and underemployment, in combination with severe restrictions on the welfare state. The environmental damage it has wrought has taken longer to emerge. Greening the neoliberal economy will not suffice to reverse these patterns of harm. The rare case where a vibrant green party has been integrated into the mainstream, as in Germany, has gone some way towards permitting public subvention of renewable energy sources but has not fundamentally challenged the neoliberal growth imperative. Germany's recent role in the austerity discipline meted out to Greece serves as a powerful illustration of this point.

Despite the fact that the SD paradigm has shaped regional, national, and international environmental governance for a generation it is increasingly obvious that climate change is becoming a potential Achilles heel of the neoliberal paradigm. While the populations of relatively affluent western nations are largely comfortable at the prospect of stark global wealth inequalities (popular sentiment against economic migration is testimony to this), deteriorating environmental conditions do generate genuine popular concern. In spite of efforts by neoliberal economists to downplay the likely economic impacts of future climate change, its actual and eventual implications are gradually seeping into the popular imagination.

Here we encounter another peculiarity of the neoliberal constitution of the social. For as long as a degree of popular support for the critique of capitalism was a widely held sentiment, a long view of political history was possible. One could speak of a slow, drawn out denouement of capitalism followed by the construction of an alternative, non-capitalist society. The collapse of state-based alternatives to capitalism, as discredited as they had become given the realities of the Soviet bloc, allowed a curiously ahistorical attitude to neoliberalism to take root. This was announced by Francis Fukuyama's (1992) widely read publication *The End of History and the Last Man*. Fukuyama's underlying thesis was that, with the collapse of the Soviet Union, the modern period of ideological struggle of the previous two centuries was over. Henceforth, liberal democracy and the open market would inexorably take root in all parts of the world. This fulfilled Hegel's pronouncement, made early in the nineteenth century, that history would come to the end with the realisation of collective human freedom. The trouble was, as Fukuyama saw it, that Hegel's original idea had been largely eclipsed by Marx's thesis of communist social revolution.

From Fukuyama's perspective, the final decade of the twentieth century served to prove Marx wrong and Hegel right. We could henceforth look forward to the relentless spread of liberal democracy, now that the spectre of world communism had been definitively vanquished. Fukuyama's neoliberal triumphalism did indeed herald a period in which traditional leftist parties across liberal democracies increasingly came to distance themselves from their origins in the international workers' movement. In fact, this was simply a further act of distancing that had characterised various European leftist parties throughout the twentieth century. Even by such historical standards there was a particularly radical demoralising of traditional socialist politics towards the end of the twentieth century. In terms of globalised neoliberalism, it was above all the so-called 'Washington Consensus' that made clear the degree of ideological convergence. This stood for a worldview of unified conviction according to which violent social revolutions would give way to non-violent, gradual conversions to the peace and prosperity offered by neoliberal capitalism.

On closer inspection, however, the hegemony of neoliberal governance has not done away with the idea of progressive world history. Rather, it has rewired the mental appreciation of what social progress amounts to. Whereas 1950s critical theory was already speaking of technological innovation having exhausted its socially useful potential in the incipient consumer society, recent decades have witnessed the virulent return of a collective consumption ecstasy. During this period, the central task of neoliberalism has been a matter of reengineering the collective social imaginary so as to grasp all phenomena through the lens of consumer choice. Such a cognitive perspective includes the permanent prospect of modifying one's choice to take advantage of improved products. While postmodernists and even modernists of a certain ilk (Frankfurt School thinkers first and foremost) had been denouncing progress in history as an illusion from the 1930s on, neoliberalism has restored collective faith in historical progressivism in the form of consumer product replacement.

Seen through this prism, however, climate change can only appear as a glorious wellspring of opportunity for green consumption. But there are obvious tensions involved in this when applied to the issue of climate change. First, climate change is not a product that can be voluntarily consumed but is rather a case of involuntary consumption (a neoliberal oxymoron). Secondly, it cannot be isolated as one distinct object of choice. Herein lies the true source of the complexity of climate change. Climate change does not name an individual phenomenon or entity within the natural environment (such as a species). Instead it refers, with relative indeterminacy, to a general condition characterising the natural environment as whole. This also accounts for the tendency of climate change to engender apocalyptic visions. As Giddens's paradox notes, climate change is politically disturbing precisely because one cannot point and say: this is what climate change looks like. Climate change can be neither localised in space nor isolated in time. Instead, it is an agonisingly slow process and affects all places on earth at once.

Reaching back to eighteenth-century aesthetics we might speak of climate change in terms of an environmental sublime. In his *A Philosophical Enquiry Into the Origin of Our Ideas of the Sublime and Beautiful*, Edmund Burke (2015 [1757]) characterised the sublime as an attitude of mind prompted by an immensely powerful presence lacking spatial determinacy. For example, the massive interior of a cathedral with a vaulted ceiling receding into darkness was designed to give rise to a sublime experience. Its immensity outdoes the capacity of human vision to take in the object perceived, thereby activating the imagination to fill in what the senses cannot. But this constructed space was itself meant as a cypher for the ultimate sublime experience of divine presence. Writing of the popularity of public executions (still a feature of urban life in eighteenth-century Europe), Burke remarks how witnessing the death of another while knowing oneself to be in a position of safety is apt to give rise to a sense of the sublime. Finally, vast natural phenomena such as the ocean give rise to sublime experience on account of their immensity and the sense of power evoked.

Both the global reach of climate change and its characterisation as a general threat to human life make it the source of a collective sublime experience. The fact that the sense of the sublime pertains to a passive spectator is also important here. The arresting of collective action is clearly a hallmark of contemporary social responses to climate change. The kind of pleasurable distress attributed by Burke to the spectators of an execution might also be assigned to the average social agent with regard to climate change. How else can the contemporary penchant for environmental disaster movies and post-apocalyptic dystopia narratives be accounted for? There is clearly a subterranean collective enjoyment at the prospect of ecological collapse. The relative safety for the average citizen across wealthy liberal democracies also bears comparison to Burke's empathetic spectators. We may well identify to some degree with the sufferings of climate refugees, but we do this from a place of relative safety.

Reflecting on what we might call the environmental sublime, it is useful to recall that neoliberal governance is not simply a matter of a dominant economic class meting out harsh social discipline. While this aspect comes to the fore in the rhetoric and reality of austerity politics, for the most part it is the ecstasies of liberated consumption that characterise the neoliberal social order. As Lyotard (1991 [1988]) noted, capitalism propagates its own sense of the sublime in the form of the consumption mantra: there will always be more. For all the talk of ecological limits since the publication of *The Limits to Growth* (Meadows et al. 1972), the success of neoliberalism can be gauged by the extent to which the mythology of limitless consumption has colonised the collective imagination.

This offers a further way of making sense of the post-historical condition realised under neoliberalism. Whereas earlier religious and secular humanist narratives posited a unique moment of collective historical transition or fruition (salvation or revolution), neoliberalism inserts historical time into the circuits of commodity obsolescence. The gradual exhausting of the natural environment by the production and circulation of commodities is not of sufficient intuitive distinctness to override the neoliberal sublime of potentially infinite consumption. There are, of course, many parallel historical narratives of progress maintained under neoliberalism (the inexorable march of universal human rights, for example), but these are ultimately subordinated to the temporalities of commodity production and consumption.

The key point here is this: climate change constitutes a social-political problem that has so far not met with anything like the kind of countervailing popular imaginary needed to respond to it. As a consequence, the more climate change is presented as a natural catastrophe, the more it is likely that neoliberal governance will be strengthened rather than weakened. The collective imagination across liberal democracies simply has no alternative conceptual scheme with enough force to counteract the neoliberal configuration. Simply proclaiming, against the false optimism of the 'green economy', that capitalism cannot rise to the challenge of climate change is important but insufficient. What is above all needed is a clear account of why climate change is insoluble under the terms set by neoliberal capitalism.

Neoliberalism has overseen the reinvention of traditional leftist politics (Giddens's 'Third Way' being simply one variant among others), thereby rendering even the modest reformism of the social democratic tradition highly implausible to many in liberal democracies. At present, climate change largely appears as an unfortunate unforeseen consequence of capitalist economic development. While the work to mitigate this consequence is deemed perfectly legitimate, the hegemony of the SD narrative ensures that this entails no fundamental challenge to the neoliberal development agenda as such. One of the problems with the attempt to found leftist opposition to capitalism on the catastrophic implications of climate change is the implication that capitalist development has only recently become 'unsustainable'. The circuits of commodity consumption tend to foreclose collective cultural memory, thereby corroding the sense of extended class history needed to inform genuinely progressive environmental politics.

Climate change should not be grasped naively as the sudden revelation that capitalist development is damaging to ecological systems. The political response to climate change should be informed by a sense of the generations-long harm capitalism has done to human and non-human systems of collective self-organisation. Across wealthy liberal democracies growing awareness of worsening social conditions brought on by massive private debt has done little to dampen the fervour of liberated personal consumption. Considering the impact of climate change on future generations, under these conditions, is likely to be reduced to the visual clichés of the average disaster movie. Just like Burke's eighteenth-century spectators, we may get a thrill out of the spectacle of another's death, but ultimately we know that everyday life will go on as normal once the body is dragged away.

Another reason to resist the short-termism induced by the catastrophe hypothesis is that it tends to promote an implausible theory of social change. Modern environmental activism over the last forty years bears witness to a continuous debate over suitable means. The default position here is that of non-violent civil disobedience. But even here one has to reflect on the nature of violence. According to a time-honoured definition of the nation state it is the only political agency that may legitimately deploy violence, either domestically or against other states (see Giddens 1985). There is, of course, a rich history of debate relating to the moral limitations of such violence in the form of just war theory.

In this context it is worthwhile noting a tendency, since the inception the 'war on terror' a decade and a half ago, to label environmental direct action 'eco-terrorism'. Direct action becomes attractive to environmentalists for obvious pragmatic reasons. If your collective goal is to protect the environment, relying on efforts to precipitate public outrage of sufficient force to bring about policy change may simply be too slow or too exhausting. Acting more directly to protect the environment can readily seem a better means to pursue one's ends. The sheer level of FBI resources dedicated to investigate and prosecute members of the Earth Liberation Front in the previous decade is illustrative of how readily the neoliberal state is willing to label such activism domestic terrorism (see Vanderheiden 2005). Ironically, this labelling shows a greater awareness of the intrinsically political character of environmentalism than that typically avowed by mainstream environmental organisations, which prefer to ground their claims with reference to the status of scientific consensus.

The invention of eco-terrorism underscores the point that the opposition between capitalism and the environment can only be resolved by a comprehensive reconfiguration of the nation state. Nothing makes it clearer that local environmental activism cannot rise to the challenge of climate change than the present readiness with which the neoliberal state is willing deploy its military and security resources against climate activists. Appeals to people power against corporate folly forget that corporations flourish under legal conditions created and maintained by national governments. According to neoliberal thinking, the task of state government amounts to facilitating the conditions of corporate innovation and market control. The reciprocal dependence of business on government is generally downplayed as merely maintaining the 'rule of law', which in our own times amounts to shoring up the legitimacy of the World Trade Organization and other legitimating enforcers of global capital.

In the context of entrenched neoliberal governance, it will never be enough to surrender the state to neoliberalism and hope that a popular environmental movement of such power and coordination arises capable of sweeping away the settled global order. Instead, the long and painstaking work of fundamentally reclaiming the nation state as social-ecological agent is unavoidable. This task has to take its initial orientation from a radically altered appreciation of socially necessary production. The current neoliberal regime has been alarming successful as severing all production imperatives from any intrinsic connection to genuine social necessity. Herein lies the source of its potential for both social and ecological destruction. Most progressive environmentalists believe that labelling this destruction a natural catastrophe is the answer. But pointing to catastrophic effects is of little value without a credible diagnosis of the political causes of climate change. Such a diagnosis should direct the task of constituting a different form of popular political agency that will not simply protest against neoliberal governance but undo it. One thing seems certain: the more climate change is invested with a sense of impending natural catastrophe the greater the chance that neoliberal patterns of production will continue into the future. For catastrophe is the very condition under which neoliberalism most readily flourishes.

CLIMATE CHANGE AND THE END OF NATURE

In recent decades the popular environmental writer Bill McKibben has published a series of books that focus on the challenge of climate change. McKibben's work has been a protracted response to the dire warnings of James Hansen, the climate scientist working for NASA who first briefed US congressional committees on the topic of dangerous global warming during the unusually hot summer of 1987. McKibben's initial intervention came in the form of this dramatic conclusion announced in his 1989 publication *The End of Nature*: 'We have deprived nature of its independence, and that is fatal to its meaning. Nature's independence *is* its meaning; without it there is nothing but us' (McKibben 2006 [1989]: 50). Drawing on the rich history of American wilderness writing, McKibben conceives of nature as something properly independent of human action. Hence, the advent of humanly caused climate change signifies the 'death of nature'. This death necessarily brings with it, as in cases of personal loss, a powerful sense of sadness and mourning. Given that climate change is for McKibben unequivocally anthropogenic the death of nature is in fact a case of murder, or at least manslaughter. As a sign of moral failure the antidote must involve choosing 'to limit ourselves voluntarily' (McKibben 2006: 182).

The tensions in McKibben's depiction of environmental crisis are not hard to identify. He begins by postulating a traditional notion of nature as utterly independent of human action, even though the crisis he articulates stems from human agency. He claims that it is 'the idea' of nature that has died, and yet draws on accumulated empirical evidence as our only proof that nature has changed radically. He insists that humanity has brought about the environmental catastrophe blindly, while calling on conscious concerted action to reverse the damage. He cautions against the quietism that stems from an insistence that humans are also part of nature, while folding humanity back into divine creation. Above all, McKibben's prose, here and elsewhere, is apocalyptic and yet his call to action politically measured and socially cautious.

McKibben's presentation of climate change is typical of the genre of environmental catastrophism. It grounds itself in a metaphysical construction of 'Nature' that can be, in certain ways, ideologically persuasive but relies on an understanding of the environment that is neither intellectually sound nor politically credible. McKibben's loyalty to the institutionalised notion of wilderness is telling in this regard. As laid down in the American Wilderness Act of 1964, wilderness refers to territory devoid of permanent human settlement and activity. As an environmental legal construct wilderness is curiously non-ecological, offering no place that might be called an *oikos* or home to humans. The narrative pathos of McKibben's *End of Nature* aligns with earlier odes to solitude in American literature. The reader is invited to feel the environmental crisis through personal pain and a wrenching sense of loss. Nature, one feels, would have been best left to an elect who might have treated it with the respect it deserves; but, alas, the serpent entered the garden, exploited the weakness and selfishness of others, and now all is (almost) lost.

In this connection it is worthwhile recalling that the original sense of the word 'apocalypse' denotes revelatory insight. With this sense in mind, to depict nature or the environment in an apocalyptic manner is never simply to indicate a factual endpoint. As McKibben's presentation amply illustrates, it is a matter of invoking a certain experience of nature. For the best part of three decades McKibben has summoned up the environmental crisis with a distinctly prophetic voice. Like one of his mentors, Wendell Berry, he mixes his message of scientific evidence going unheeded with a religious sense of abusing an earth entrusted to humanity's stewardship. This religious register may be appealing or repellent depending on the sensibilities of the reader, but the key point is how it characterises the appropriate social-political response to climate change. If the origin of the problem is ethical-religious waywardness, then the appropriate cure is collective moral reform.

This diagnosis is quite at odds with the analysis offered in this book, which asserts that climate change is an impersonal, systemic symptom of neoliberal capitalism. On this view, we cannot combat climate change through moral reform. Instead, it should be seen in connection with the growth imperative that characterises global capitalism. In his activism with 350.org McKibben has notably sharpened the political edge of his environmentalism, appealing directly for more enlightened policy-making as a matter of urgency. However, as his more recent book *Deep Economy* (McKibben 2007) made clear, he instinctively prefers to depoliticise climate change by appealing to a vision of local environmental autonomy he feels almost all Americans, regardless of political inclination, could endorse. This tendency stems from McKibben's underlying conviction that climate change is a problem of blind human greed and lack of foresight with respect to future generations. Once again, this points to a moral problem with a moral solution, rather than seeing climate change as the result of a certain, historically grounded, system of material production and social organisation.

Given the apocalyptic presentation of climate change the most surprising feature of McKibben's environmental analysis is the tameness of his envisioned social reform. In this he seems to be simultaneously relaying two messages that are hard if not impossible to reconcile: first, that climate change is the greatest and most intractable challenge humanity has ever faced; and, secondly, that we can meet this challenge through a process of non-violent grassroots environmentalism. On the first count, the catastrophic narrative summons up a social-political problem so vast and portentous that only a leviathan-like social agency could hope to be equal to it. And yet, the solution of environmental localism offered in *Deep Economy*, with illustrations of local independent radio stations and small-scale energy generation, does not convince following this depiction of the scale and depth of the environmental crisis we face.

The organisation that McKibben has more recently helped found and promote, 350.org (named after the parts per million concentration of atmospheric carbon dioxide generally deemed safe by climate scientists), constitutes the social movement embodying this perspective. While 350.org certainly deserves praise for attempting to create a global network of progressive climate activists, these efforts are seriously vitiated by its general stance of political neutrality. This neutrality is not uncommon among those climate activists who appeal to environmental localism. The mantra here is something like: 'Let's just get it done!' The UK-based Transition movement is another prominent example of this approach (see Hopkins 2013). There is a kind of naive populism at work here, according to which local people know best and so should be left, as much as possible, to conduct their own experiments in alternative ecological life. This neglects the fact, highlighted in the previous chapter, that climate change is much too extensive and radical an environmental condition to be tackled by largely uncoordinated local environmental social agents.

The extent to which 350.org is willing to politicise climate change essentially comes down to appealing to politicians to do the right thing. But this reduction of politics to ethics – in common with the belief in local autonomy rather than centralised governmental action – is itself a hallmark of neoliberal politics. Just as patterns of production and consumption come down to the capabilities of the well-informed consumer, so too political action is thought to be ultimately reducible to the individual politician's cultivation and application of practical wisdom. It is up to the political class, therefore, to resist the lures of corporations who are not doing the right environmental thing. This ethical reduction essentially places all agents, regardless of institutional context and social standing, on the same level. To this is added the naive conviction that there is no inherent reason why capitalist production need be environmentally destructive. If enlightened citizens and corporate leaders demonstrate the moral superiority of their approach to climate change, then politicians will inevitably convince the fossil fuel companies to leave their reserves in the ground. The obvious problem with this approach is that the foundational capitalist drive for profit is by no means a function of personal greed. Neoliberal capitalism can neither be understood nor overturned by appealing to moral psychological analysis.

RECLAIMING ENVIRONMENTAL POLITICS

McKibben's end of nature thesis thus strikes the wrong note from the outset. As indicated, his conception of nature is clearly aligned with the American tradition of wilderness writing. This tradition was established in the nineteenth century, a period of western thought and culture deeply coloured by the various European romanticisms (chiefly British and German) that rediscovered nature just as industrialisation was taking hold in the last decades of the eighteenth century. While it was a stretch even in the eighteenth century to identify anything like a genuine British wilderness, this did not prevent romantic poets and artists from offering its sophisticated audience a vestige of wild nature. Partly this constituted a kind of elegy to vanishing modes of rural existence as a consequence of the rapid urbanisation of the population (see Williams 1973).

This socially conservative aspect of romanticism (finding its German equivalent in Johann Gottfried Herder's anti-Enlightenment celebration of the local as opposed to the universal) was complemented by a celebration of the dignity of the common agricultural worker. The second wave of British romanticism in the nineteenth century intensified the sense of trammelled wilderness by transferring concern to the socially marginalised poetic individual. In the works of Byron, for instance, nature could be poetically portrayed as a refuge from social inauthenticity and artificiality. The romantic flight to the Mediterranean prepared the popular imagination for mass tourism in the second half of the twentieth century. Childe Harold's existentially driven excursions can be seen as pioneering 'eco-tourism'.

McKibben's interpretation of climate change as the end of nature represents the supposedly final act in this romantic tragedy. Henceforth, the effects of fallen humanity are omnipresent and unspoiled nature is no longer available even as a marginal refuge. Nature has finally become fully fictionalised in the sense of something only available to us in tales of past experience. Once again, the problem here is the non-specificity of the fatal agency prosecuted for the death of nature. Humanity in general does not amount to a determinate political agent. Portraying the end of nature without detailed historical specificity allows environmental sentiment to merge seamlessly with the neoliberal narrative that claims history is over. It is telling, on this score, that McKibben's end of nature is pronounced at the same time as Fukuyama's end of history. Both narratives trade in the pathos of irreversibility, the former in the mode of elegy, the later in the mode of euphoria. The elegiac stands out of history and rhetorically asks: were we not fated to kill nature all along? Posed in this way, the religious undertones of McKibben's environmentalism become prominent and decisive. Long before he embarked on the activism of 350.org McKibben had made his key decision: no politics can be sufficient to bring nature back from the dead.

Given McKibben's status as an influential figurehead of progressive environmentalism in the United States it is worthwhile reflecting further on his political position. Beyond the centrality of fossil fuels and an ideology of wealth accumulation McKibben offers no specificity regarding the economic causes of climate change. The pivotal role of neoliberal economics is deliberately passed over in silence. This is chiefly because McKibben believes that 'politicising' climate change will prevent the formation of a sufficiently broad popular movement to fight it. But this misses the key point: climate change is no less politically constituted than nature itself. Herein resides McKibben's chief theoretical and practical error. There has never been such a thing as nature considered independent of human thought and action. Nature could not die because it has never existed as the kind of thing McKibben takes it to be.

In other words, nature is 'socially constructed', but not in the sense of being a mere mental projection. Rather the point is, following Latour, that nature grasped as some sort of superobject has never existed. Reflecting on the origins of science Aristotle referred to his Greek precursors as 'nature thinkers'. The Greek term for nature (*phusis*) is used in ancient texts as a synonym for universe (*kosmos*). In other words, 'nature' simply means everything that exists. But the term has a further meaning, as when we refer to 'nature' as the inherent essence of a certain thing. In the first line of his *Metaphysics* Aristotle (1999) asserts that every human desires 'by nature' (*phusei*) to know. He means by this that knowing is the essential and unique activity of the human being. In his *Politics* Aristotle (2009) considers whether political organisation (social organisation within the ancient city-state) is a function of human nature or merely a conventional arrangement for mutual convenience. He famously concludes that political organisation is not simply an economic expedient but rather an expression of human nature itself.

Returning to contemporary environmental catastrophism, it becomes evident how the two distinct meanings of 'nature' already found in ancient Greek thought are conflated. The 'end' of nature really amounts to destroying the nature (essential character) of Nature (everything that is). But this narrative readily dissolves into absurdity. First, because it is an empirical impossibility that human beings could exist in the absence of certain ecological conditions (i.e. Nature cannot be absent as long as some forms of life remain). Secondly, at least since the advent of organised agricultural practice around 10,000 years ago, human action has significantly transformed natural systems (that is, the nature of Nature has been constantly changed by human agency). Consequently, nature is at an end for McKibben because he implicitly adopts the romantic ideology of nature as a kind of sublime object transcending human influence.

The confusions and inadequacies of McKibben's discourse of environmental catastrophe all stem from the lack of a credible materialist, historically grounded understanding of social agency. While physically climate change is a direct consequence of using up carbon-based fuel sources, historically speaking the fossil fuel-based economy has a specific social-political constitution. If, following the development of this global economy, nature is truly dead then it becomes unclear what environmental activism would be fighting for. A less rather than more dead nature? A dead nature treated to morally superior funeral arrangements? Perhaps we can best understand the role of nature here on analogy with the Hollywood action hero, who seems to die multiple times during the course of the narrative, but miraculously survives to fight again. Similarly, each time climate catastrophe is announced the proximity of nature to its final death scene is seemingly increased. Just as each episode in the narrative of the romantic hero prefigures his death, so too in the genre of environmental catastrophism nature can only ever appear through a superimposed death mask. In this light, McKibben's version of catastrophism simply serves as an illustration of a much broader tendency, one more recently shared by Klein (2014) in her depiction of the titanic clash between capitalism and the environment.

The point of the critique advanced here, to repeat, is not to deny the existence or gravity of climate change. Rather, it is to insist that announcing a crisis without a cogent political analysis of the specific social agencies that have produced it will simply allow the perpetuation of the neoliberal order. Climate change is, properly grasped, a core 'contradiction' of neoliberal capitalism. Climate change is a global ecological condition engendered by an economic paradigm that is so successful that any political alternative has been rendered virtually unthinkable. Neoliberalism also functions, borrowing Klein's (2007) term, as a 'shock doctrine'; that is, as a political regime that feeds off conditions of social crisis. For this reason, those who earnestly wish to combat climate change should eschew catastrophism in favour of the long and arduous task of bringing about the transition to an alternate mode of political organisation along socialist lines. To those who resort to the language of catastrophe to grasp the social challenge of climate change, we must insist on posing the question: what, precisely, does this invocation of catastrophe reveal as the necessary transformation beyond the neoliberal order? In posing this question, we must always count on the fact that neoliberalism is already far advanced in working out its own version of what lies 'beyond' climate change.

THE NEOLIBERAL RESPONSE TO CLIMATE CRISIS

The argument advanced in this chapter involves two core contentions. First, catastrophism naturalises climate change and so masks its specific political causes; second, framing climate change as a natural catastrophe is highly likely to perpetuate the very political order (neoliberal capitalism) within which the present global environmental situation has arisen. The attractiveness of climate catastrophism derives from many sources. It has been demonstrated using the example of McKibben's thesis of the end of nature that these sources are chiefly religious, literary and moral in kind. An Old Testament invocation of nature (plagues, famine, and so forth) deploys the pathetic fallacy, according to which natural phenomena are signs of divine moral agency.

When these framings of nature are put to work within the context of neoliberal governance, however, they are immediately translated into further grounds for intensified capitalist development. On the one hand, the precariousness of ecological resources prompts the neoliberal state to further efforts aimed at ensuring access to natural resources. This has the effect of exacerbating the power imbalance between developed and developing nations. On the other hand, social consciousness of risk is heightened, prompting voters to stick with what they know rather than pursue the risky strategy of fundamentally changing political course. In other words, climate catastrophism promotes the preservation and intensification of the political status quo.

In *Justice, Nature, and the Geography of Difference* David Harvey (1996) remarks: '[D]ifferent conceptions of nature get evoked for quite different political and substantive purposes within the overall flow of conflictual social action' (Harvey 1996: 173–4). In other words, building a political platform on the premise of fighting for or defending nature (or the climate) really begs the question: what kind of nature? Given the status accorded to science, we might think this question could be unambiguously resolved by responding: nature as described by the natural sciences. The problem with this, as our discussion of Latour made clear, is that it rests on the spurious assumption that modern scientific inquiry is not subject to all kinds of social, cultural, and political determinations. Even if the political neutrality of science could be defended it is clear that the desire for purely 'science-led' policy-making is bound to be frustrated in the context of liberal democracies.

Plato's *Republic* (2008) illustrates this point beautifully. Contemporary philosopher-kings would precisely be science-led technocrats dedicated to rigorously impartial rule in the interests of the whole community. The problem with this political vision within a modern context is obvious: it departs completely from the democratic project. The role of 'the people', the principle of popular rule, would have no place in a polity run exclusively by the scientifically enlightened elite. But the more relevant naivety of science-led politics in our own times relates to the fact that scientific activity is clearly subordinate to the imperative of neoliberal development. Even the most banal reconstruction of modern science would readily show the pivotal role of technological applications aimed at enhancing capitalist productive efficiency. While neoliberal ideology prefers the mythology of the individual entrepreneurial inventor as the real scientific hero, 'normal science' (see Kuhn 2012) is characterised by largely anonymous research groupings with the ultimate goal of producing marketable commodities. Given the scientific consensus on climate change, the basic problem for the neoliberal order is: how can climate science be translated into enhanced profit-making? The general answer to this question is already apparent: in the form of enhanced security technology and infrastructure.

It is important to note how the environmental (that is, natural resource) security agenda of neoliberalism does not amount to an attempt to prevent or control climate change, but merely to turn a profit on it. In *Uneven Development* Smith (2008) asserts a key distinction that helps to clarify this point: 'The production of nature should not be confused with *control* over nature [. . .] The production of nature is not somehow the completion of mastery over it, but something qualitatively quite different' (Smith 2008: 87–8). Smith points out that capitalist exploitation of nature can in fact never achieve plausible mastery of nature, nor is it particularly interested in such mastery. Instead, the task is to extract profit from the exploitation of natural resources.

This distinction between the capitalist production of and control over nature is important because it highlights the following point: climate change is an issue for neoliberal governance only insofar as it presents new opportunities for and challenges to profit making. In other words, the social harm caused by climate change is not intrinsically problematic within the neoliberal paradigm. After all, it is clear enough that a pervasive sense of social precariousness is one of the key political symptoms of this mode of governance. If climate change were to induce levels of social disturbance that endangered capitalist productivity in general this would indeed represent a problem for neoliberalism. But there is currently, for all the climate catastrophism in circulation, little prospect of this happening at a systemic global level.

The US Department of Defense (DoD) *Climate Change Adaptation Roadmap* (US DoD 2014) sets out the envisaged national and international security implications. It is worthwhile quoting a passage at length to give a sense of this specific framing of the problem:

> The changing climate will affect operating environments and may aggravate existing or trigger new risks to U.S. interests. For example, sea level rise may impact the execution of amphibious landings; changing temperatures and lengthened seasons could impact operation timing windows; and increased frequency of extreme weather could impact overflight possibility as well as intelligence, surveillance and reconnaissance capability. The opening of formerly-frozen Arctic sea lanes will increase the need for the Department to monitor events, safeguard freedom of navigation, and ensure stability in this resource-rich area. Maintaining stability within and among other nations is an important means of avoiding full-scale military conflicts [. . .] These developments could undermine already-fragile governments that are unable to respond effectively or challenge currently-stable governments, as well as increasing competition and tension between countries vying for limited resources. These gaps in governance can create an avenue for extremist ideologies and conditions that foster terrorism.

A striking feature of the DoD's assessment of climate change is how it is perceived to be a threat to domestic security only to the extent that it engenders political instability in other countries. Envisaged domestic social harm is framed not as a political issue but merely as necessitating greater use of military resources for disaster relief. The presentation of developing nations as badly governed or even ungovernable is a characteristic feature of neoliberal ideology. Of course, 'bad governance' in neoliberal terms signifies in truth low penetration of global capital into a national territory.

The economic imperative of the neoliberal climate change security agenda becomes quite transparent where the document refers to the need to 'ensure stability' in the 'resource-rich area' of the Arctic. In November 2015 Shell declared that it would not, for the time being, explore for further oil reserves off the Alaska coast despite winning permission to do so from the US Federal government (see Macalister 2015). Hailed by climate activists as a political victory for environmentalists, this decision is, more plausibly, a reflection of the low global price for crude oil. This illustrates once again Smith's contention that the capitalist production of nature is directed by the basic impulse to turn a short-term profit, rather than some generalised will to exploit nature for its own sake.

The DoD assessment might lead one to think that the United States is not, historically, the nation state chiefly responsible for causing climate change and instead merely interested in safeguarding the world from its threats. Given the social harms of climate change enumerated in the passage – 'impairing access to food and water, damaging infrastructure, spreading disease, uprooting and displacing large numbers of people, compelling mass migration' – it is striking that the option of tackling the root causes of climate change is nowhere mentioned. This follows the pattern of naturalising climate change within the catastrophe narrative. Accordingly, rather than attempt to explain the political-economic context in which global climate change has come to the fore, the global environmental situation is taken as a given fact. The only genuine problem thus remains how to keep capitalism on track on the terms desired by the most powerful neoliberal states.

Claiming that the United States is merely interested in exploiting climate change in order to consolidate neoliberal rule might strike many as unfair and one-sided. Despite its notorious refusal to be a signatory of the Kyoto Protocol (the 1992 intergovernmental framework to reduce carbon emissions), the United States does have an impressive history of environmental policy-making aimed at protecting air quality, endangered species and critical natural habitat. Furthermore, President Obama has led earnest efforts to promote renewable energy, most prominently solar, and has mandated higher Federal standards of fuel efficiency for vehicles. Despite this superficial green agenda, however, the Obama administration is more fundamentally driven by the security framing of climate change. A May 2015 Whitehouse document titled 'The National Security Implications of Climate Change' summarises the situation as follows:

> Climate change is an urgent and growing threat to U.S. national security, contributing to increased weather extremes which worsen refugee flows and conflicts over basic resources like food and water. The national security implications of climate change reach far beyond U.S. coastlines, further threatening already fragile regions of the world. Increased sea levels and storm surges threaten coastal regions, infrastructure, and property. A changing climate will act as an accelerant of instability around the world, exacerbating tensions related to water scarcity and food shortages, natural resource competition, underdevelopment, and overpopulation. (Whitehouse 2015: 8)

In tone and substance this assessment of the geopolitical implications of climate change is identical to the DoD stance. Once again, there is no acknowledgement that the historical causes of climate change intrinsically relate to the neoliberalising of the global economy, a process led by the United States. The double standards of the Whitehouse assessment of climate change emerge most readily in its presentation of the Arctic region. The disappearance of sea ice and the consequent navigability of Arctic shipping lanes have significant economic implications. Chief among these is the opportunity to exploit formerly inaccessible fossil fuel reserves.

Were it not for the state of Alaska, the United States would struggle to rationalise its political interest in the Arctic. As it stands, this region is likely to become a key theatre of international rivalries precipitated by climate change. As such, it will be a critical bellwether of how global politics are likely to develop as the effects of climate change sharpen all over the globe. One thing is already abundantly clear: the hegemonic neoliberal states will not hesitate to use military means to maintain and expand the current political-economic regime. The capitalist drive for access to fossil fuel resources has never shown much concern for the social harm caused to populations who happen to be situated where these resources are to be found. Keeping with this pattern, the Arctic peoples are highly likely to join the growing masses displaced directly or indirectly by the effects of climate change.

ENVIRONMENTAL JUSTICE AS DISTRIBUTIVE POLITICAL ECOLOGY

Through the lens of neoliberal governance climate change is not a condition politically determined as such, bur rather a natural phenomenon used to justify an intensification of neoliberal rule. The most powerful neoliberal states, above all the United States, view climate change, first and foremost, as a security problem. But 'security' here amounts to protecting the social-economic conditions under which neoliberal economic and social condition can flourish. Such security in no way precludes and in fact guarantees the creation of sacrifice zones, where massive social harm is accepted as the inevitable cost of capitalist exploitation of natural resources. For example, it seems clear that the interests of people living in the Arctic region will be sacrificed for the sake of realising its enhanced economic potential in the form of newly accessible fossil fuels and time-saving shipping routes.

But if this response to climate change is a political construction rather than a natural necessity what are the credible political alternatives? Within the environmental movement a prominent narrative of environmental justice (EJ) has arisen in recent decades. In many ways, EJ follows a familiar pattern of social justice. Originally, social justice was simply a political-ethical platform for fairer wealth distribution. Environmentally considered, social justice includes the distribution of ecological benefits and burdens. For example, a metropolitan government pursuing environmental justice would ensure equal access to parks while avoiding concentrations of environmental toxins in any one area or neighbourhood.

The EJ movement originally took the form of struggles against such concentrations in poor neighbourhoods populated predominantly by communities of colour (see Bullard 1993, 2000, 2003).

In this way there is an important historical connection between EJ and the fight for racial justice following the Civil Rights movement. The EJ paradigm makes clear that the value of the natural environment is never a politically neutral given and is in fact the consequence of hardened patterns of unequal distribution. As Engels (1988 [1872]) noted of the industrialised urban populations of nineteenth-century England, the physical environment is just as much a construction of capitalism as the commodities produced and circulated under this economic regime.

The shortcoming, however, of the EJ approach when it comes to climate change is that it is simply not radical enough to disrupt the neoliberal logic. Despite continuous criticism, the spurious 'trickle down' social narrative promoted by neoliberalism since the 1980s has lodged in the collective unconscious across liberal democracies. This acts to maintain popular faith in the potential of neoliberal economics to deliver enhanced social justice. In effect, the EJ paradigm does not fundamentally question how value is constructed under neoliberalism. Rather, it merely pushes for a fairer distribution of such value. The very fact that the growth of the EJ paradigm has occurred in tandem with the consolidation of neoliberal hegemony should give us pause. While class antagonism in the Marxist tradition works according to the assumption of capitalism's intrinsically harmful exploitation of nature (in the first instance, human labour), nothing in the EJ approach insists on a fundamental contradiction between capitalist economics and environmental welfare. But it is crucial to recognise the existence of just such a contradiction. This means that realising environmental justice in reality entails undoing neoliberal governance.

Neoliberalism's promotion of a natural catastrophe narrative in relation to climate change is complemented by a more fundamental projection of social utopianism. That utopianism is rarely recognised in its banal garb of 'economic development'. Nevertheless, positing endless capitalist growth is undoubtedly utopian in character. Like Burke's sublime object, neoliberal capitalism inspires awe precisely because progressive growth has no discernable limits. Distributive justice paradigms, including EJ, strive to make a limited, beautiful object out of the neoliberal sublime. But all such efforts at reform are bound, ultimately, to fail. As noted in the previous chapter, the economic growth metaphor has been so thoroughly naturalised that opposing it seems akin to proclaiming the earth to be flat. In a curious way, any purely redistributive counter to capitalism simply strengthens the hold of the neoliberal utopia on the collective imagination. After all, the very success of the SD paradigm rests on the principal article of faith that increased economic development is the only way to enable social and environmental well-being in the poorer countries of the world. It is both difficult and yet absurdly simple to observe, by contrast, that more accumulation of wealth in no way ensures fairer distribution.

In Chapter 4 I will tackle the core problem of neoliberal capitalism's characterisation of labour as 'human resources'. It is equally important to challenge the neoliberal concept of the environment in terms of 'natural resources'. In many ways, this is what the modern environmental movement has aimed at all along. Looking at nature exclusively as a resource for human exploitation clearly promotes a blinkered appreciation of the natural environment. If an old-growth forest, for example, is simply viewed as a potential source of timber for construction and fuel it can be readily exploited without regard to its much more comprehensive ecological significance. Traditional limits on exploitation were underpinned by the relative geographical fixity of the exploiting human population. These limits are precisely removed by the dramatically intensified and generalised economic flows ushered in by industrialisation, thereby allowing devastating exploitation in one region to benefit others.

While natural resource scarcity is used as a justificatory premise for the existence of capitalist economic management (the 'tragedy of the commons' notion), such scarcity is in fact a construction or effect of capitalism itself. This framing of environmental concern was demonstrated above in connection with the national security interpretation of climate change. As we saw, climate change is not presented as a symptom of neoliberalism but rather as a natural condition that neoliberalism is best placed to deal with. In effect, the prescription of neoliberalism is always the same: take more of the poison in order to be cured of the disease.

The sense that a perfectly functioning market could only have ecologically benign results is an article of faith best displayed by the academic field of environmental economics (see Turner et al. 1994). As a subfield of neoclassical economics, environmental economics has flourished in a post-Brundtland world. As opposed to ecological economics, which defends so-called 'strong' sustainability (non-declining natural capital), environmental economics occupies the position of 'weak' sustainability, according to which natural capital may be diminished in exchange for non-natural capital (other, human-made resources of equal or great value for human well-being). The underlying idea of environmental economics is that the ability to calculate the entire economic costs of environmental harm (sometimes called 'full cost accounting') will bring about ecologically enlightened environmental action to avoid or at least mitigate such harm.

It is not hard to see how environmental economics has grown to elaborate on the stance of the Brundtland (United Nations 1987) SD agenda. As with Brundtland, the general approach of environmental economics rests on the unwarranted conviction that environmental harm is not an intrinsic feature of the neoliberal management of the economy. Considered the unwitting agent of climate change, such management is supposedly open to the kind of enlightened reform promised by the calculations offered by environmental economics. Such economics could thus function as the means to realise the aspirations of environmental justice. A well-managed neoliberal economy could thereby coincide with an ideally just society. This natural resource management perspective is generally so uncritically assumed by policy makers and public discourse alike that it can be difficult to see the problem.

Put most simply, natural resources are material potentialities to satisfy human needs. But human needs, in turn, are never merely a matter of biological fact but also socially and culturally constituted. When environmental justice configures nature politically as a set of scarce resources and pursues greater equity in distribution and decision-making, the fact that such resources are universally capable of satisfying social needs is largely taken for granted. But this is tantamount to assuming that there is a common discourse of nature shared by all partners to the justice discussion. But this is not something that can be assumed either theoretically or practically. In practice, the configuration of nature in terms of resources almost inevitably favours dominant powers within any discursive framework. Nature as resource is always nature subject to a certain 'interested' interpretation.

Just as Latour insists that the facts established by the natural sciences are never value-free, so too natural resources should always be viewed as a political construction. This was a key insight arrived at by Marx (see Sitton 2010: 95–122), who insisted that it was naive and ultimately fruitless to say that the market price of capitalist commodities is not a socially valid measure. On the contrary, such a price is the exclusive and unerring measure of value under capitalism. The critical point of the Marxist critique, however, is that there are fundamentally different ways of constituting value. Smith (2008) brings this key Marxist insight to bear in his analysis of the production of nature:

> Socialism [. . .] is the arena of struggle to develop real social control over the production of nature. Early in his life, Marx pictured communism as the 'genuine resolution of the conflict between men and nature' [. . .] The struggle for socialism is the struggle for social control to determine what is and is not socially necessary. Ultimately it is the struggle to control what is and is not value. Under capitalism, this is a judgment made in the market, one which presents itself as a natural result. Socialism is the struggle to judge necessity according not to the market but to human need, according not to exchange-value and profit, but to use-value. (Smith 2008: 89)

Here Smith underscores his cardinal point that politics is ultimately about social control over the production of nature rather than control or mastery over nature itself. As Foucault's (2008) analysis makes clear, neoliberalism is not to be grasped as a form of politics in the narrow sense of a certain style of policy-making or genre of economic discourse. Rather, it is a comprehensive social-political paradigm that sees the market as the sole domain in which value is generated. At the time of Foucault's lectures in 1979 (Foucault 2008), climate change had not become a pivotal social issue. But we can apply his concept of neoliberalism to this current issue and say that under neoliberalism climate change will be viewed as a problem of maximising global natural resource efficiency. As the exclusive source and context of all efficiency calculations, neoliberal environmental-ism will inevitably look to the market to generate all credible ecological solutions. Counter-market proposals will accordingly be cast either as ungrounded utopianism or as ideological aber-rations to be framed as threats to security (that is, as terrorism of one kind or another).

Following Smith's production of nature idea, however, it is more accurate to see any and every mode of governance, including neoliberalism, in terms of utopian construction. Producing nature means, among other things, that nature is configured according to a specific social imaginary. In other words, nature is always ideologically overdetermined. What is currently striking about the general discussion of climate change is how readily it is being used as an occasion to reassert the unavoidable necessity of neoliberal governance. At every turn we see the effects of climate change interpreted as symptoms of a global condition calling for more potent applications of neoliberal management. The picture pre-sented by climate activists in certain quarters, according to which climate change is a 'game changer' and so forth, seems wilfully ignorant of the extent to which the climate 'crisis' is being used to justify not less but more potent and far-reaching applications of neoliberal governance. But this should come as no surprise. For all its glib talk of innovation and progress, neoliberal governance is permanently concerned to ensure nothing radically new enters the political arena. In this sense, Fukuyama's 'end of history' has proven to be eerily prescient. The intrinsic terror of neoliberalism consists in its constant erasure or absorption of all alternative con-structions of the social imaginary.

POLITICS AFTER NATURE

The fundamental issue analysed in this book is whether neoliberal governance can resolve the complex of social challenges brought about by climate change. My contention is that such governance is not only incapable of doing this, but that these problems will continue to be exacerbated under neoliberalism. The core pre-sumption of neoliberalism, according to which central planning stifles economic innovation, efficiency, and growth, will need to be overturned to allow for the kind of intergovernmental action climate change calls for. But this is not simply a case of returning to old-style command economies. As originally envisaged by Marx, what ultimately needs to change is something fundamental about the construction of social value. In Smith's (2008) words, this is 'the struggle to judge necessity according not to the market but to human need' (Smith 2008: 89).

The underlying, but mostly unrecognised, aim of modern environmentalism is such a critique of market value. Whether this is in the form of an 'anthropocentric' concern for the physical environment as it affects human health or issues from the 'ecocentric' motive of protecting species diversity and ecosystem integrity, the more general environmental concern is the flourishing of living beings. This political ecology in some ways makes for a strange politics but in others harks back to an older conception of political organisation. Ancient Greek thought tended to think of politics in precisely this way, namely as organising the collective for the good of all. The original sense of 'economics', after all, was household management.

Modern environmentalism and ecology appeal to this collective good as it relates to all forms of life. Such a truly inclusive concept of politics may seem fanciful in light of the rival interests within any given human community. How, it will be asked, can we hope to reconcile the interests of all living beings when the human collective remains so fractious? Here Latour can again be of help to us. For his part he insists that the key question of political ecology should be: how do we welcome non-humans into politics? For Latour, the central concern of ecology should not be construed as a question of rendering distribution of environmental goods more equitable. It is rather about instituting a credible and genuine 'politics of things' (Latour 1999: 203), whether those things are natural or non-natural.

The traditional paradigms of ethical theory will be variously challenged by this characterisation of environmental politics. If values are qualities of things only in virtue of human evaluation, then rival environmental values can be reduced to competing sources of evaluation. In other words, natural things are merely the innocent screens onto which human values are projected. But Latour rejects this construction of the problem in principle.

It is central to what he sees as the grand project of modernity to constitute a cardinal separation of powers between society and nature, politics and science. But the manifold challenges that constitute the 'environmental crisis' precisely call for the dismantling of these separations: 'Political representation of non-humans seems not only plausible now but necessary, when the notion would have seemed ludicrous or indecent not long ago' (Latour 1999: 202).

A genuinely ecological polity would, therefore, reshape traditional democratic politics in a quite radical fashion. What Latour is pointing to is the kind of political transformation that responds to the ecological mantra 'everything is connected'. Clearly, this political interpretation of environmentalism is very far from the current reality of liberal democratic politics, where environmental policy is at best humoured but certainly never allowed to trump the neoliberal imperative of economic growth. Latour holds that contemporary society and politics remain in thrall to modernist hubris according to which human subjects are considered the sole political agents. Latour's hunch that the so-called wicked problem of climate change indicates a certain unravelling of the modernist constitution of politics seems to me sound. But I remain less sanguine that the political upshot of this unravelling will produce anything of genuinely democratic value without the kind of long-term preparation for social reorganisation once found within the labour movement. Latour's hope that political ecology will finally allow us to awake from the dogmatic slumber of modernity will not occur without protracted social struggle against the principal agencies of neoliberal governance. As indicated, this necessarily entails reconstituting national government in a counter-neoliberal fashion.

A more straightforward way to arrive at the kind of post-Nature (that is, beyond nature understood as intrinsically independent of humanity) politics envisaged by Latour is to acknowledge that climate change has brought human dependence upon natural systems into focus to an unprecedented degree. The neoliberal paradigm has two principal ways of responding to this: first, through faith in the innovative potential of technological development; and, second, through the belief that raising educational standards in developing countries will lead to a long-term decline in the global population. On the first count, the naivety stems from seeing technological progress as invariably socially benign; the second attempts to gloss over the close connections between capitalism's need for cheap labour and its inexorable drive for profit extraction.

Against the first presumption, one salient example relates to the coercive manner in which food technology is foisted on developing countries in the name of 'food security' (see Clapp 2006). It has been difficult and in some cases impossible for the developing world to maintain more traditional agricultural practices in the face of economic bullying on the part of giant agribusiness and its sponsor states. In addition, developing countries can be readily scapegoated as the main cause of environmental problems due to burgeoning populations. What gets lost in this picture is the economic pressure exerted on such countries to urbanise at hitherto unknown rates. This, in turn, is driven by the relentless need of capitalism for cheap pools of manufacturing labour to supply goods to the more affluent consumer nations. Urbanisation also disrupts patterns of rural land management, thereby facilitating agricultural land purchase and management by large multinational agribusinesses.

According to the neoliberal utopia, eventually all nation states will enjoy the fruits of a developed economy, at which point stabilisation of the global population will have been achieved. In the actual world, the systemic effects of neoliberal governance that have led inexorably to massive internal displacement and outward economic migration make the original optimism of the Washington Consensus looks increasingly delusional. But, again, we should not be surprised at this. The underlying motive of capitalism has always been intensified profit making and this motive must necessarily sweep aside all genuinely ecological considerations.

Whereas the original concept of market-based value formulated by the likes of Adam Smith frames the market as an impersonal social context in which the base matter of individual greed is transformed into the gold of collective flourishing, neoliberal capitalism has in fact generated a condition of endemic human misery and ecological degradation. Rather than Smith's sympathetic eighteenth-century burgers across wealthy liberal democracies we

find citizen-consumers beset by social and economic precariousness. Discontent with neoliberal globalisation, rather than leading
to critique of neoliberal policies, more often generates a populist
politics scapegoating social welfare recipients and other vulnerable groups such as migrants. Despite the fact that the Great Recession offered a momentary glimpse of a neoliberal economic system
built on sand, neither governments nor social movements across
liberal democracies proved capable of bringing about a fundamental change of course. Instead, the cure for what ails neoliberal
politics was, as ever, taken to be insufficient neoliberal economic
discipline. Neoliberal governance has from the beginning operated
according to a 'no alternative' logic. Today's austerity politics are
simply the latest act in this drama.

But let us return to our central issue of climate change and
recall that widespread consciousness of the 'environmental crisis'
became a key social issue just as neoliberal governance was consolidating its hold over society. Today, environmentalism within
liberal democracies often looks more like a special interest of
the pampered middle class than a radical political movement of
the dispossessed. In the context of increasingly severe neoliberal
social disciplining, those who find themselves struggling to meet
everyday material needs are unlikely to see environmental issues
as paramount. Melting icecaps are not going to be a lively concern when someone is out of work and struggling to maintain a
minimal standard of living.

The reality is, of course, that there is no shortage of wealth
across liberal democracies. It is axiomatic that, under capitalist conditions, developed economies become wealthier year on
year. Post Great Recession the verdict among left-leaning thinkers appears to be that the problem with neoliberal economics is growing inequality (see Piketty 2014). I believe this is a
mistaken diagnosis. One has to delve deeper in order to reach
something more fundamental. This something, as I have been
arguing in this chapter, is the neoliberal construction of value.
Achieving better distribution of the spoils of capitalist development, while undoubtedly welcome in the short term, will not
fundamentally undermine the causes of climate change. What
is true of the world economy more generally holds within the
more narrow confines of the developed economies: the goal

of greater wealth distribution is the false promise that allows neoliberalism to remain socially attractive despite its manifest shortcomings. Fairer distribution of capitalist wealth, whether achieved through the mechanisms of the state or otherwise, is not a solution for climate change.

This is another reason why the programme of environmental localism is a political dead-end. Allowing local communities to control their own energy production and so forth simply allows central government to devote more energy to the neoliberalisation of the global economy. In terms of social policy, the state can continue to insist that suitably energised citizens should possess the wherewithal to provide for themselves, as opposed to those who remain dependent on state welfare. In other words, environmental localism is a largely unwitting example of the very social entrepreneurialism promoted by neoliberal governance.

Conceiving political ecology in Latour's (2004) terms, by contrast, means grasping environmentalism as a call to extend the political project of democracy in a truly radical manner. The enfranchisement of the non-human, it is fair to say, is still very far from anything like a popular sentiment in today's liberal democracies. Those who, wherever they place themselves on the political spectrum, present climate change as an impending natural catastrophe will find the prospect of such a radical change in political paradigm an annoying distraction from the real work of staving off or preparing for ecological collapse. But the radical nature of the problem of climate change calls for just such a radical transformation in politics. As I argue throughout this book, this transformation should not be thought of in terms of the traditional Marxist revolutionary moment. While climate change is, I believe, a pre-eminent 'contradiction' of capitalism, is has not appeared at a historical juncture where an enlightened proletariat is waiting to seize power.

Across liberal democracies genuine class-consciousness in the Marxist sense has been largely eclipsed by the neoliberal social imaginary. The social pathologies of the consumer society, originally diagnosed by critical thinkers in the 1950s, have now become normalised through integration into an uncritical horizon of popular common sense. Hence the widespread message that the only manner of environmental agency left to the average citizens is to exercise consumer power by 'buying green'. But consumption is not a viable framework for political agency. As Smith's notion of the production of nature indicates, the only veritable source of political power stems from having a role in the co-production of value. We, the end-users, have little to no say in how our consumer goods are produced further 'upstream'.

Part of the heritage of early socialism is the idea that communities themselves should have such control over production. This is arguably what survives today in the various experiments in environmental localism. But the socialist visions that accompanied the advent of modern industrialisation cannot be simply revived in the current era of neoliberal capitalism. Following Latour's (2004) thinking, what is needed instead is some synthesis of the radical environmental idea of a politics beyond human subjects with the socialist conception of a social transformation geared towards the 'use value' of all living beings.

In order to work towards this we must stop seeing climate change as a natural catastrophe. Catastrophic consciousness is not amenable to the long, painstaking project of transforming politics in the manner indicated. The immediate task in achieving this transformation is the reconstruction of class-consciousness, that is, being able to see shaping the material environment of life as a matter collective co-production. In other words, we need to take up again the original political promise of the modern democratic transformation of society. Only now, 'we, the people' must be radically expanded to 'we, who inhabit the earth'. This convocation is the real political challenge posed by climate change.

CHAPTER 3

Sustainable development as neoliberal environmentalism

NEOLIBERALISM AND SUSTAINABLE DEVELOPMENT

The basic contention of this chapter is that the apparently politically neutral notion of sustainable development (SD) amounts to neoliberal environmentalism. It is not by chance that the problem of climate change has arisen precisely in the period dominated by the SD approach. As noted in the previous chapter, leading neoliberal states present climate change within a broader political construction of security. Within this construction even the most positive relations between the physical environment and human flourishing (in such areas as nutrition, and so forth) are cast as contexts of mutual struggle and risk calculation. The paradigm of sustainable development, while it admits of many variants, at base represents environmentalism under the aegis of neoliberalism. It follows from this that the way in which climate change is predominantly represented means that the genuine social problems it entails cannot in any real sense be tackled. In short: climate change is a salient symptom of neoliberal environmentalism. But it is a symptom pointing to a fundamental social pathology for which neoliberalism can offer no credible therapy.

In saying this I do not, of course, intend to deny that the historical origins of climate change go back to the early Industrial Revolution of the mid-eighteenth century. While there were reactions to the process of early industrialisation in politics and literature, there was no pervasive sense that the basic functions of the physical environment as a whole were imperilled. The recent adoption of the term 'Anthropocene' (see Steffen et al. 2007; Davis 2010; Zalasiewicz et al. 2011), whatever its analytical usefulness, is powerfully indicative of the fact that this sense does indeed characterise the present age. My core argument in this chapter is thus twofold: first, that it is precisely under global neoliberal governance that the problem of climate change could emerge; and, second, that the SD terms on which this problem are set ensure that the underlying issues of social organisation will not be tackled.

The fact that all major transnational petrochemical companies explicitly align their business mission with sustainable development provides direct support for the claim that SD amounts to neoliberal environmentalism. A few examples by way of illustration:

> ExxonMobil Fuels & Lubricants (F&L) recognizes the importance of addressing sustainability in today's global marketplace – balancing economic growth, social development, and environmental protection, so that future generations are not compromised by actions taken today. (Exxon Mobil 2015)

> Sustainability at Shell touches on all areas of our operations. We aim to deliver the energy needed for a growing population in a responsible way – respecting people, their safety and the environment. Sustainability is essential to the longevity of our business and our role as a member of society. (Shell 2016)

> We believe that the best way for BP to achieve sustainable success is by acting in the long-term interests of our shareholders, our partners and society. We aim to create long-term value for our investors and benefits for the communities and societies in which we operate. (BP 2016)

Given the fact that the burning of fossil fuels is the basic physical cause of climate change, one could be forgiven for attributing quite pathological levels of self-deception to these corporate claims. How did it become possible for the principal agents responsible for climate change to pose as guardians of environmental stewardship? It is an understandable reaction of much environmental activism to demonise these corporations and their executives.

But this approach does little or nothing to reveal the fundamental nature of neoliberal governance as it relates to the environment. Neoliberalism is not reducible to social dominance of the capitalist corporate class, but also crucially entails a distinct form of the collective imaginary in which entrepreneurialism becomes the dominant social norm. In this way, neoliberal reality is as much in the heads of individual social agents as it is out there in the world of corporate-led production. Indeed, the very environmental groups that seek to save the environment from predatory capitalism all too often resort to the green consumption model to fund their campaigns and lobbying efforts.

The fact that neoliberalism is not simply a state of mind limited to corporate capitalists but rather pervades the entire social milieu of contemporary liberal democracies means that the rest of society cannot be accurately presented as mere passive victims of neoliberal governance. It is nevertheless true that neoliberalism aims at a fundamental pacification of social dissent, a process that undoubtedly pertains to modern environmentalism. Considered broadly, the original platform of the environmental movement from the 1960s and 1970s was a comprehensive critique of industrialised society combined with various practical attempts to live beyond the industrial paradigm. With the beginning of the global neoliberal transition in the mid 1970s it was initially seen as necessary to discredit or co-opt all forms of anti-capitalist critique, whether the intellectual dissent disseminated largely within higher education or the practical programmes of reform organised by various civil society protest movements (see Harvey 2005: 39–63).

What is striking about the neoliberal conversion is its syncretic mode of operation: neoliberalism seldom removes opposition by explicitly opposing it but rather recasts it in a seemingly more socially acceptable form. Those opponents who remain recalcitrant can eventually be cast as unreasonable or dangerous radicals, extremists, or, ultimately, terrorists. The ultimate goal is to generate a collective conviction, a sort of common ideological baseline, according to which neoliberal solutions represent the only acceptable manner of reform-minded modernisation. Neoliberalism has proven so successful in thus reshaping the social imaginary because it is capable of offering a kind of something-for-everybody smorgasbord of reform. For free market advocates it promises the end of state-led economic planning; for traditional liberals it calls forth a vibrant civil society of non-profit organisations; and for left-leaning progressives it facilitates a comprehensive programme of civil rights for marginalised groups. It would be foolish to contend that, on all these fronts, nothing of genuine worth had been realised by neoliberalisation. But such acknowledgement should not prevent us from identifying the basic drive of neoliberalism to naturalise the capitalist market and subsume all social relations under it. Such naturalisation allows social and environmental harm on a huge scale to be presented as inevitable products of economic-political organisation.

THE ORIGINS OF NEOLIBERALISM

In *A Brief History of Neoliberalism* David Harvey (2005) characterises neoliberalism as a political-economic order that links human flourishing with private property rights and free markets. According to the neoliberal position, he remarks, 'social good will be maximised by maximising the reach and frequency of market transactions, and [neoliberalism] seeks to bring all human action into the domain of the market' (Harvey 2005: 3). While the origins of neoliberal theory can be traced back to the Freiburg 'Ordo-liberalism' school of the 1920s and 1930s and the work of the Austrian Friedrich von Hayek in the 1940s (see Foucault 2008; Hayek 2011 [1960]), it was ultimately the favourable reception of Milton Friedman's work from the 1960s and 1970s (see Friedman 2002 [1962]) that forged the hegemonic variant of neoliberalism. This neoliberalism, beginning around the time of the global oil crisis in 1973, has displaced the 'embedded' liberalism of the 1950s and 1960s whereby an implicit stability pact between capital and organised labour was maintained within each liberal democratic state.

Harvey (2005) offers a useful shortlist of key transitional factors: a shift from manufacturing to finance; significant consolidation of CEO power; the rise of biotechnology and information technology as key economic domains; and the prominence of international institutions such as the International Monetary Fund, the World Bank, and the World Trade Organization. Using indicators such as the reduction of real wages and the exponential concentration of wealth at the top over the last three decades, Harvey sees the rise of neoliberalism in classic Marxist terms as a clear expression of class power (Harvey 2005: 31). While initial experiments in neoliberalism were carried out by force of American geopolitical power in South America in the 1970s, the Reagan–Thatcher period of the 1980s required more subtle means to re-engineer mature liberal democracies. Harvey again points to a number of factors: key theoretical contributions such as Robert Nozick's (2013 [1974]) *Anarchy, State, and Utopia* and Milton and Rose Friedman's *Free to Choose* (1990 [1980]); significant expansion of the American Chamber of Commerce; and the establishment of rightwing think tanks such as the Heritage Foundation, the Hoover Institute, and the American Enterprise Institute (Harvey 2005: 43–4). Once the influence of neoliberal thinking across the spectrum of traditional party politics was consolidated in capitalist countries, neoliberalism as a theoretical position could increasingly be passed off as unassailable common sense in university business schools and economics departments in the US and elsewhere.

Identifying key factors in the historical development of neoliberal ideas and practices can help to dispel its aura of inevitability. Like climate change, the term neoliberalism names an exceedingly complex set of social-political phenomena. My overall argument is that climate change is a problem that is presented but not solvable under the terms of neoliberal governance. The SD paradigm is the specious solution that neoliberalism offers for all environmental problems, including the problem of climate change. Unlike the term neoliberalism, which until recently was not in widespread use, the concept of sustainable development has become a ubiquitous banality, something almost all environmentalists support without troubling to clarify was is meant by sustainable. Captured in succinct terms, sustainable development denotes a process whereby neoliberal capitalism is presented as the sole means of attaining social-ecological flourishing. In its most direct articulation as environmental economics it presents the environment as a subset of the economy. Following the original impetus of Ordoliberalism, the SD paradigm projects an ideal case of perfect market operation where all material goods are given their proper social valuation.

Relative to the radical environmentalism of the 1960s and 1970s the neoliberal position promotes a social vision whereby corporate and citizen interests may seamlessly integrate without fundamental opposition. In this way, the SD paradigm projects perpetual peace between capitalism and environmentalism. This logic reaches its apogee in the kind of public relations exercises of large petrochemical corporations noted at the start of this chapter. In promotional videos testifying to their environmental concern, these corporations combine emotive music and panoramas of unspoiled nature with images of contented employees and customers, as though to say: 'Look, all your worst fears about our impact on the environment are baseless. Human flourishing is our prime concern'.

Where the advertising veneer wears thin an alternate tactic of ethical contrition provides a further avenue for neoliberal pacification. In this case, rather than insisting that all is well with the world, we are confronted by confession and the need to change behaviour in some fairly dramatic way. An example of this would be American political leadership admitting the country is 'addicted to oil'. This manages to cast a century-long element of advanced industrialisation as some kind of psychological complex, a sort of neurotic behavioural pattern that could be corrected with suitable therapy. The beauty of this tactic is that it implicates everyone caught up in the neoliberal economy. If society is addicted to oil, then petrochemical corporations are simply servicing that, admittedly ill-advised, desire. Equally, if we decide to kick the habit then business will accommodate the change.

Finally, neoliberal environmentalism grasps entrepreneurial innovation as the essential means for realising sustainable development. Whether to meet consumer expectations or because it ultimately constitutes good business sense, sustainable development is presented as the corporate model of the future. We find a similar internalisation of corporate-led sustainability when it comes to consumption patterns by citizen-consumers. Buying green works in large part by carrying a social marker of superior ethical responsibility. The fly in this particular ointment, of course, is that products are subject to a green mark-up in price, meaning that poorer consumers are obliged to make environmentally inferior purchasing choices. As recent developments in organic food production have illustrated, however, at certain economies of scale 'sustainable' goods can be made attainable to most consumers in wealthy liberal democracies. Ultimately, however, just as there are 'failed states', there will inevitably be a social remainder in liberal democracies regrettably lacking the spending power to join the sustainability revolution (see Guthman 2004).

THE BRUNDTLAND REPORT ON SUSTAINABLE DEVELOPMENT

The concept of sustainable development within policy-making has a fairly clear history. Although the term 'sustainable development' had been used sporadically before then, the notion began to assume prominence with the 1987 publication of the Brundtland Report, *Our Common Future*, produced under the auspices of the United Nations World Commission on Environment and Development (WCED). The report offered the following, thereafter canonical, definition: 'Sustainable development is development that meets the needs of the present without compromising the ability of future generations to meet their own needs' (United Nations 1987). The problem that had led to the report's existence was essentially twofold: on the one hand, how to get the most economically developed nations to cooperate in the arena of environmental policy; on the other hand, how to facilitate industrialisation in developing nations without overburdening the environment. While fundamental political critique with regard to capitalist development is carefully avoided, Brundtland's elaboration of the concept of SD does allow for preferential treatment of developing nations' basic needs over the developed world's non-basic needs. Since the publication of the Brundtland Report an entire genre of environmental ethical literature has amassed arguing about the merits of this preference.

Given real-world economic developments since the mid-1980s, however, the significance of any ethical rationalisation of

SD is in practical terms negligible. The SD programme is essentially economic-political rather than moral in nature. While moral philosophers have debated for decades what models of equitable distribution best conform to the ideal of SD, intergovernmental policy discussions have been characterised by protracted stand-offs between rival economic blocs. Even seemingly counter-neoliberal economic policies such as massive state subsidisation of agriculture by the European Union and the United States have been maintained in the name of such rivalry. Developing countries, where they could, have responded to the SD framework by taking developed nations to task as the chief historical agents of climate change and other environmental problems. It is fair to say that, three decades after its official establishment as a key strategic policy framework, sustainable development has achieved neither of its original aims: it has not led to a credible shared programme of environmental protection among developed nations, nor has it facilitated environmentally benign economic development elsewhere. A key indicator of both is the failure to tackle climate change.

But we should not be surprised by this failure. Seen for what it actually is, namely environmental neoliberalism, SD has performed exactly as intended. Despite waves of anti-capitalist protest and the occasional instance of organised 'eco-terrorism', sustainable development has become at once a universal corporate mantra and a social code of self-discipline across wealthy liberal democracies. A virtually perfect alignment of capitalist business and civil society (the more general goal of neoliberalism) has thereby been achieved in the environmental sphere. Beginning as a marginal protest movement in the 1960s modern environmentalism has achieved true maturity in the era of neoliberalism. Now everyone can say with equanimity: I want to be sustainable. But while, subjectively, being sustainable clearly means different things to different people, objectively SD really amounts to one thing: generating profit with a good environmental conscience. One might therefore reduce the actual SD argument to: if you support capitalist development, you support genuine environmentalism. The tautological nature of this formula is revealed by its reversibility: it equally holds that being a genuine environmentalist entails supporting capitalist development.

Meanwhile steady-state or negative economic growth models of political ecology first proposed in the 1970s are dismissed as out of touch with basic economic realities; or else as proponents of a developed world chauvinism wanting to deny the benefits of capitalist development to the rest of the world; or, alternatively, of wanting to force humanity back to the Stone Age. The SD platform has to this extent been hugely successful: it has comprehensively transformed modern environmentalism from a counter-cultural critique of capitalist development into a business-friendly framework for relentless neoliberal social reform. It is in the historic context of the success of SD that climate change has arisen as a global environmental problem that puts in relief both the nature and limits of neoliberal governance. In terms of social organisation adequate responses to climate change call for a kind of cooperative governance of the commons that is necessarily ruled out by neoliberal capitalism. The solutions to climate change proposed within the SD paradigm will not fundamentally touch the social problems implied. Rather than the forward-looking environmental entrepreneurialism favoured under SD, the real solutions must be sought by drawing on the rich history of socialist alternatives to the capitalist organisation of production.

FOUCAULT ON NEOLIBERALISM

Neoliberalism constitutes, to echo what Jean-Paul Sartre (1963) could say of Marxism in the late 1950s, the 'unsurpassable horizon' of our age. To repeat the basic claim of this chapter: sustainable development is nothing other than neoliberalism as applied to the environment. The very incorporation of environmental concern into neoliberalism might seem on the surface of things rather surprising. Why not leave environmentalism to wither through the neglect and marginalisation accorded to other 1960s protest movements? The answer stems from neoliberalism's peculiar interest in 'the environment' considered more generally. Foucault (2008) hinted at this interest but did not develop the point in his 1979 lectures on neoliberalism. He says there:

> [. . .] if you define the object of economic analysis as the set
> of systemic responses to the variables of the environment, then
> you can see the possibility of integrating within economics a set
> of techniques, those called behavioural techniques, which are
> currently in fashion in the United States. (Foucault 2008: 270)

Foucault is here alluding to the behaviourist paradigm in modern psychology as represented by the work of B. F. Skinner (see
Skinner 1976). While the environment in question is clearly the
human rather than natural environment, Foucault is intimating
that neoliberalism ultimately aims at the constitution of a kind
of totalised regime of environmental testing. This means, among
other things, that neoliberal governance is not to be grasped
according to the traditional notion of ideology, Marx's replacement for the Hegelian *Weltanschauung* (worldview). Neoliberalism is not simply a cognitive lens through which to grasp reality,
but, more fundamentally, a mode of material reality or total
environment. In its own manner neoliberalism brings about the
end of nature by making the human and natural environment
one integrated field for sustainable development.

SD is customarily analysed into three mutually supporting
pillars relating to the nested or overlapping domains of the economic, environmental, and social. Much of the critical literature
on the SD paradigm amounts to the charge that the environmental and social pillars are subordinated to the economic. While
this critique is undoubtedly well founded it fails to illuminate the
grounding of SD in neoliberalism. Neoliberalism, as suggested,
offers a unified concept of the environment such that the distin-
guished domains of sustainability are initially collapsed into a
sort of homogeneous plane, constituting a concept of environment indifferent to the social-natural distinction. Conceptually,
this is what allows climate change to be predominantly presented
as a natural, rather than a social-political, condition.

In Foucault's analysis the construction of a comprehensive concept of environment is achieved through the fundamental notion
of enterprise. Enterprise here is not simply the idea that the social-
economic context offers opportunities; it is instead a general
invitation to each social agent to regard themselves as an inex-
haustible opportunity for self-realisation. Unlike Marxist social

critique, neoliberalism focuses on the worker as entrepreneurial agent rather than on the product of work as an object of exchange:

> In neo-liberalism – and it does not hide this; it proclaims it – there is also a theory of *homo oeconomicus*, but he is not at all the partner of exchange. *Homo oeconomicus* is an entrepreneur, an entrepreneur of himself. This is true to the extent that, in practice, the stake in all neo-liberal analyses is the replacement every time of *homo oeconomicus* as partner of exchange with a *homo oeconomicus* as entrepreneur of himself, being for himself his own capital, being for himself his own producer, being for himself the source of [his] earnings. (Foucault 2008: 226)

As Foucault also points out, this neoliberal reconceptualisation of labour undercuts the Marxist critique according to which the worker under capitalist conditions is alienated from their living labour through its reduction to exchange value. In the push and pull between workers' wage levels and commodity prices capitalist pressure to increase profit margins inexorably leads to wages representing a progressively smaller proportion of the overall value of production. Neoliberal analysis of the same set of phenomena shifts the perspective, from one that looks at work from the outside according to exchange mechanisms to one that looks out at the world as seen through the eyes of the worker. The neoliberal worker is, unlike the Marxist version, not a victim of alienating market mechanisms but rather an agent capable of realising a variety of marketable abilities. This latter perspective applies not only to production but also consumption. In fact, neoliberalism elides the Marxist distinction between production and consumption in this way:

> The man of consumption, insofar as he consumes, is a producer. What does he produce? Well, quite simply, he produces his own satisfaction. And we should think of consumption as an enterprise activity by which the individual, precisely on the basis of the capital he has at his disposal, will produce something that will be his own satisfaction. (Foucault 2008: 226)

Foucault is particularly firm on this point: common sociological critiques of the 'consumer society', according to which capitalist social existence is depicted as torn asunder into irreconcilable roles of producer and consumer, are worthless in the context of neoliberalism. Three and a half decades later it is no doubt easier to acknowledge this point, now that the consumption paradigm has colonised virtually every domain of social existence. The same process is clearly in evidence in the case of the conversion of environmental politics into sustainable development. Following Foucault's analysis, we can say that political problems relating to the natural environment have been folded into a more comprehensive environment of entrepreneurial consumption. To this extent, sustainability is both a bureaucratic management framework (resource-saving rationalisation) and a paradigm of consumer self-realisation (environmental entrepreneurialism).

Returning to Foucault's allusion to the role of American behaviourism within the development of neoliberalism, it is necessary to underscore the point that we are not here confronted by an ideology in the traditional Marxist sense. If that were the case, then the present historical juncture would have to be characterised in terms of near universal 'false consciousness'. But this term, like the broader notion of alienation, would amount to portraying neoliberalism as a kind of involuntary mass delusion. As indicated earlier, however, the point is that neoliberalism more generally and SD more particularly are not reducible to a way of seeing things; they equally, and more fundamentally, amount to a certain fashioning of the environment itself.

TEST CASE: NEOLIBERAL EDUCATION REFORM

The current neoliberalisation of public education serves to illustrate this point. This process is particularly prominent in the charter school movement in the United States and the wave of academy and 'free' school creation in the United Kingdom. What gets mostly lost in public debate (regarding the merits or demerits of decentralising management, establishing standardised testing regimes, school performance tables, and so forth), is the more profound transformation of the common social environment or public sphere. Fundamentally, neoliberalism cannot tolerate the existence of a commons that is not subject to market logic and discipline. The imposition of this discipline draws on the traditional liberal suspicion of state management and combines this with an insistence on bureaucratic accountability (output metrics touted as operational transparency). Rhetorically, this double-sided process is presented as returning public schools to the public via the individual choice mechanism. This rhetoric is, in turn, founded on a basic tenet according to which the social environment is to be managed in order precisely to maximise consumer choice. The rationale for this tenet is, as Foucault also observed in his lectures, essentially utilitarian in character: the overriding norm of social existence is individual satisfaction (the maximisation of happiness or balance of pleasure over pain, in classical utilitarian language).

One can observe this neoliberalising process at a relatively early stage with regard to the current changes underway within public education. Rather than a social provision necessary for fostering a vibrant democracy, public education is being recast as a form of individual, private self-investment. Accordingly, it is enough if enhanced educational opportunities are provided. Failure to avail oneself of these opportunities is a sign of weak entrepreneurial spirit and the resultant negative utility will allow individuals and families to see the virtue of making superior choices in this domain. The original impetus for state-funded education, by contrast, stemmed from recognition of endemic inequalities that could not be corrected and in fact only worsened through exposure to market pressures. Prior to the creation of state-funded and managed systems in the early twentieth century, education for the poor was provided by enlightened capitalists in conjunction with humanitarian or religious charity. In typical fashion, neoliberal education reform does not repudiate the goal of ameliorating inequality but rather claims it has means superior to the state for doing so.

According to proponents of neoliberal education reform, it is the traditional, state-run model that has perpetuated social marginalisation by failing to hold all learners to sufficiently high standards and not offering parents real educational choices for their children. Due to the current hybrid neoliberal solution (publicly funded, privately managed), two mechanisms are simultaneously deployed that largely work against each other: on the one hand, school choice is decoupled from residential location creating thereby diversification of consumer options; on the other hand, the imposition of increasingly time-consuming standardised testing actually creates pedagogical homogenisation with less room for experimentation. This combination arises largely because the provision of testing facilities is currently the most lucrative enterprise open to private business within the public education sector. Thus, while maximisation of choice is the message neoliberal school reform relays to educational consumers, the overriding imperative of profit-making through testing provision tends to neutralise any diversity achieved.

The greater loss, however, entailed in the neoliberalisation of public education is further erosion of the public sphere across liberal democracies. In this context it is not insignificant that public education is one of the last areas of mass employment that still largely enjoys union protection. By contrast, as of 2012 only 7 per cent of all US charter schools had a unionised workforce (Rebarber and Zgainer 2014: 3), thereby leaving them vulnerable to lower pay and working conditions compared to their traditional public school counterparts. Neoliberalising public education is thus grounded in the more basic notion of the neoliberal economic agent as, in Foucault's (2008) words, 'entrepreneur of himself, being for himself his own capital, being for himself his own producer, being for himself the source of [his] earnings' (Foucault 2008: 226). In his lectures Foucault explicitly refers to the neoliberal conceptualisation of education within the context of human capital considered more generally. In the following remarks he links formal education with a much more comprehensive environment of investment in the child:

> Time spent, care given, as well as the parents' education . . . in short, the set of cultural stimuli received by the child, will all contribute to the formation of those elements that can make up a human capital. This means that we thus arrive at a whole environmental analysis, as the Americans say, of the child's life which it will be possible to calculate, and to a certain extent quantify, or at any rate measure, in terms of the possibilities of investment in human capital. (Foucault 2008: 229–30)

Foucault's description of the neoliberal learning environment highlights two aspects: first, the parental or caregiver role of prudent investor in the child; and, secondly, the calculability of likely return on investment. The same analysis may be applied, as Foucault goes on to say, to other domains, such as public health, domestic migration, and the criminal justice system. The common thread in the neoliberal logic applied to all such domains is the concept of individual investment and risk management. Surveying the political landscape from the second decade of the twenty-first century the upshot of the neoliberal reform in these various arenas of public life is evident.

To take higher education as a case in point, the reality of state-funded college education has been destroyed in many liberal democracies over the last three decades. The oft-repeated rationale for this is none other than the personal investor analogy. The rhetorical question here is: why should society pay to enhance an individual's personal economic capacity? Either a college education provides this enhancement or it does not. If it does, the individual will be repaid through higher wages; if it does not, the individual will be induced to make a better investment choice next time around. Either way, the role of good government is to allow individuals to be effective entrepreneurs, that is, take responsibility for risk calculations connected to personal investment decisions. Government should strive to see that there is no transparently obvious fraud involved in this entrepreneurial environment, but it should certainly not undertake the role of managing services itself.

This neoliberal line of thinking has become such second nature among liberal democratic policy-makers and public discourse alike that it is often hard to recall that there was ever an alternative way of seeing things. Often those who lead protest movements against neoliberalisation unwittingly deploy a similar logic. As with the environmental movement's attraction to localism, so the idea of self-determination can be very enticing to educational progressives. When local schools seem to offer a poor choice, ambitious parents in gentrifying neighbourhoods will be prone to search for alternatives to the traditional publicly funded school; when publicly funded hospitals seem under-resourced, it is tempting to go private where that option exists.

Building on the example of public education, we can say more generally that neoliberalism has reshaped the collective imaginary so that every act, from selecting a life partner to choosing a variety of coffee, is seen as a matter of smart personal investment. As popularised social critique has it, discontent with neoliberalism often stems from a subjective sense of drowning in options in all areas of life from the most mundane to most profound. The central issue for the present analysis is how SD presents climate change in neoliberal terms. The answer, at least schematically, can be very simply put: climate change is presented as a matter of speculative entrepreneurialism. In other words, the neoliberal economy must configure climate change as an opportunity for profit-making. To some, this observation may seem grotesque, horrifying, or ludicrous to various degrees. In his lectures Foucault commented on the laughter provoked by his presentation of choosing a spouse as a matter of wise investment. 'I am not saying this as a joke,' he remarked, 'it is simply a form of thought or a form of problematic that is currently being elaborated' (Foucault 2008: 228). What struck Foucault at the end of the 1970s as an experimental line of thought now constitutes hardened social reality across contemporary liberal democracies.

CLIMATE CHANGE AND THE ROLE OF THE STATE

The subject line of a 1991 memo attributed to Larry Summers, then Chief Economist at the World Bank, runs: '"Dirty" Industries: Just between you and me, shouldn't the World Bank be encouraging MORE migration of the dirty industries to the LDCs [Least Developed Countries]?' (Enwegbara 2001). The story behind the origins of the memo and the intentions behind it have since developed into a full-blown conspiracy theory. Whether Summers was the actual author and whether it was meant to be ironic does not concern me here. Instead, the question of significance is whether the environmental economic reasoning expressed in the memo accurately represents the actual functioning of the neoliberal order.

In the first section of this chapter the dominant framing of environmental concern by sustainable development was identified as neoliberal environmentalism. My basic critical claim is that neo-liberal governance in its SD formulation remains fundamentally blind to the problems of social organisation bound up with the issue of climate change. Consequently, sustainable development is not the framework of environmentally sound management it is almost universally touted to be. Equally, SD is not capable of reform such that it could be rendered environmentally benign, as many progressives currently contend. For example, the development of ecological economics (see Costanza 1992) that seeks to critique the subordination of environmental sustainability to economic sustainability (found in more standard environmental economics) represents a well-meaning reform of SD that does not amount to a fundamental challenge to the neoliberal paradigm. Neoliberal capitalism will never learn to internalise harm to the environment to the extent that this overrides the basic impetus towards profit-making. The best that neoliberal environmentalism can offer is ways of restructuring environmental protection that accord with its entrepreneurial model. What neoliberalism is constitutionally incapable of, we should never tire of reminding ourselves, is leaving intact or creating a commons not subject to market discipline. Such discipline invariably compromises the ability of an environmental commons to allow the flourishing of all that exists within it.

The idea articulated in the Summers memo, whatever its original intentions, is that environmental burdens should be situated where they entail the lowest economic cost. In other words, it is a matter of managing environmental externalities in the most 'rational', market-based manner. As the text of the memo elaborates, when it comes to things such as atmospheric pollution, for example by airborne carcinogens, economic costs rises in line with the standard of living. The example given is the risk of dying of prostate cancer. In poorer countries significantly fewer people will live long enough to develop such cancer. Thus, the presence of atmospheric carcinogens where most will not live to develop this cancer implies lower costs. This harsh Malthusian logic has long been cause for criticism among environmental justice movements fighting against concentrations of environmental toxins in poorer neighbourhoods. The geographic scale may vary but the economic logic is the same: situating environmental burdens among the poor will minimise costs. In a similar vein, Giddens (2011) connects climate change and global uneven development:

> The bulk of the emissions causing climate change have been generated by the industrial countries, yet its impact will be felt most strongly in the poorer regions of the world. A basic sense of justice should help drive attempts to reduce that impact, but there are more selfish reasons for the more affluent countries to help the more deprived too. Extreme poverty is potentially very destabilizing indeed in world society. The level of risk it produces for the more favoured countries and regions, even if global warming didn't exist, would still be formidable. Among other harmful effects, poverty is one of the main influences leading to population growth; population pressures ease as countries become richer. (Giddens 2011: 298)

These same arguments could be used for promoting distributive justice when addressing stark economic inequality within wealthy countries, yet neoliberal austerity discipline roundly rejects the soundness of this logic. The basic problem with Giddens's approach is his unwarranted belief in the possibility of a benign version of neoliberalism, a version originally cast in the form of 'trickle-down' economics. As Giddens's own account of intergovernmental climate negotiations makes abundantly clear, divisions among blocs of economically developed as opposed to developing nations have meant that even modest targets for global reductions of greenhouse gas emissions have not been achieved. Among wealthier nations there are certainly cases, generally in northern Europe, where growth in renewable energy production in recent decades has been impressive. But the more powerful imperative of economic growth has ensured that overall reductions in carbon emissions have been modest even in these cases.

Carbon markets, seen as a potential corporate self-regulation solution for climate change over the last two decades, represents the false hope of neoliberal environmentalism (see Spash 2010). The idea is that industry is allocated limited levels of free pollution and charged for anything above the allocation. In reality, both countries and companies gamed the carbon trading systems, overestimating the baseline of necessary pollution. To date, Giddens (2011) reports, it is hard to ascertain whether carbon trading has had any positive net effect in curbing the build-up of greenhouse gases. Leaving business to regulate itself is, of course, an axiom of neoliberal governance. Where carbon reduction schemes are at least tolerated by business leaders, the idea is that corporations themselves should design and implement them. Such corporate self-regulation stands opposed in principle to such measures as state-imposed mandatory limits on unwanted ecological externalities. Old-style environmental policy that sets absolute limits to pollution (for example, the US Clean Air Act) is considered out of step with more sophisticated governance methods that allow corporate-based innovation to provide environmental solutions. The glaring practical problem with this preference is that it has manifestly not helped to curb the global carbon emissions that are driving climate change.

Assuming that neoliberal governance will one day wake up to the challenge of climate change and start acting in a genuinely prudential manner amounts to ungrounded optimism when the underlying imperatives of neoliberal governance are recognised. As the course of international negotiations over the last two decades makes abundantly clear, climate change has sharpened divides between developed and developing nations, with no prospect of drawing them closer for the sake of the global environmental commons. Giddens (2011) argues that the only way to make credible headway in reducing carbon pollution is through state policy that takes the initiative rather than relying simply on corporate self-regulation:

> *The state must make interventions into markets to institutionalize 'the polluter pays' principle,* thereby ensuring that markets work in favour of climate change policy, rather than against it. In almost all developed countries at the moment environmental costs remain largely externalized. I am dubious about how effective carbon markets as such will be, but there is a great deal that can be done to introduce full cost pricing, and therefore to allow market forces to become centred upon promoting environmental benefits. Government should act to reduce 'negative externalities' – situations in which environmental costs are not brought into the marketplace – in order that markets can work to environmental ends. (Giddens 2011: 95–6)

Giddens adds to this a further principle of state intervention, whereby governments actively work against corporate interests seeking to undermine legislation to counter climate change. In light of the failure of economic deregulation, he adds, there 'has now to be a return to greater state interventionism' (Giddens 2011: 99).

Giddens explicitly acknowledges that the intellectual foundations of neoliberalism, as established by a thinker such as Friedrich von Hayek, involve a deep-rooted hostility towards economic state planning. Hayek (2011 [1960]) argued that the free market allows liberal democratic society to develop in an open, piecemeal, and organic fashion. State-led planning, by contrast, involves forcing the manifold social interactions of the market onto the Procrustean bed of national economic plans. Rather than adjusting to reality on the ground in line with the empirical method in science, state economic planning follows the rationalist tradition of innate knowledge. The problem with this, for Hayek, is the state's knowledge is eminently fallible and, more often than not, economically ruinous.

In typical 'Third Way' style Giddens (2011) envisages a middle way between freewheeling neoliberalism and old-style state planning: 'In terms of the economy, ways will have to be found to introduce regulation without crippling that sense of adventure and entrepreneurialism upon which a successful response to climate change will also depend' (Giddens 2011: 99). There are also, Giddens notes, problems of scale and linkage connected to state planning. The spirit of environmental localism that inhabits much climate change activism, as noted in previous chapters, tends to view centralised planning with suspicion. Equally, one nation's environmental planning, no matter how radical and robust, will not suffice to tackle climate change without concordance with other countries.

The most problematic element of Giddens's (2011) call for more state intervention as the principal way to combat climate change is his, distinctly neoliberal, faith in the possibility of environmentally benign corporate action. It is all a question, in his words, of governments acting 'together with enlightened corporate leaders' (Giddens 2011: 96). This idea expresses a key article of neoliberal faith. Belief in progressive social change through enlightened corporate leadership is a prominent feature of the neoliberal paradigm. The basic idea is that competence in business connotes political decision-making competence. This belief expresses itself most directly when political candidates tout their business acumen. The conceit, of course, is that governing a nation is analogous to running a business. It does not take much, in fact, to undermine the analogy. A business executive is paid to maximise profits for shareholders within a competitive framework. A politician, by contrast, is expected to rise above factional interest and govern in the interests of the entire electorate or 'commonweal'.

The more democratic politics has been taken into the neoliberal fold, the more the constituency (the enfranchised electorate) has been dissolved into competitive interests. This tendency is perhaps most apparent in the case of welfare policy and the endlessly repeated rhetorical divide between 'hard-working' citizens and the feckless unemployed or underemployed who rely on 'government handouts'. Social stigma attached to the latter has been extremely effective in dampening the traditional socialist critique of dehumanising working conditions. At the apex of the neoliberal social pyramid, of course, is the corporate superstar, who has routed their competitors in the pitiless neoliberal struggle for existence. Corporate entrepreneurialism involves, then, a curious process of alchemy, transforming profit-seeking competitiveness into a capacity for fostering general social flourishing. This blurring of private enterprise and public governance is a distinctive feature of neoliberalism. It is one of the chief notions that must be dispelled, rather than strengthened, if the state is truly to become the kind of environmental political agent envisaged by Giddens.

THE PROSPECT OF FUTURE CLIMATE WARS

In the previous chapter the militarisation of state responses to climate change was briefly considered. While the neoliberal paradigm generally frowns upon governmental action that constitutes long-term economic planning of a Keynesian sort, the state's role in using force to defend the rule of law in domestic and overseas contexts is not similarly challenged. Certain contemporary commentators have long highlighted a linkage between neoliberal governance and a state's tendency to use coercive force both internally and externally in direct defiance of the democratic ideal of popular self-rule (see Chomsky 2005). In general, no such force is needed, as neoliberal governance enacts social discipline in large measure through the rhetorical creation of constant social crisis. We should view the dominant framing of climate change against this well-established historical tendency to smother all impulses towards popular self-governance. It is important, however, not to jump to the conclusion that a credible response to this tendency is offered by the kinds of environmental localism critiqued earlier. Such localism rests on the erroneous belief that the corporate state can be left to its own devices while local communities are left to pursue progressive environmental projects. Neoliberal governance, it should be clear by now, is not characterised by this sort of benign neglect of civil society, at whatever scale. Where the impulses of localised community threaten to undermine the precedence of the market-centred perspective, they will be subject to unrelenting opposition at the hands of the neoliberal order of things.

Under neoliberal governance a motley parade of crises is constantly constructed, but the underlying effect is quite uniform and crucially involves stifling political pressure to transform conditions of work for the majority of workers. In this sense the extreme austerity politics that a country such as Greece has experienced over the last five years is characteristic of neoliberalism's tendency to generate a more generalised state of emergency. From this perspective, the spectre of climate change was always likely to become integrated into neoliberal crisis management. The most cogent neoliberal argument offered for combatting climate change is that the economic costs of doing nothing will rapidly mount in the course of this century as the global economy is disrupted.

On the one hand, this seems like sound common sense. If you can fix something relatively cheaply now, why wait until costs will be much higher in the future? On the other hand, neoliberalism is arguably more internally consistent when it looks to climate change as a global economic opportunity. As long as markets in climate change resilience prove lucrative enough, there may in fact be no sound economic argument for mitigating it. Globally speaking, the negative effects of climate change will be very unevenly distributed. Recalling the logic of the Summers memo, the most likely long-term response to climate change under neoliberal governance will be attempts to bolster dominance over developing economies by exploiting their disproportionate exposure to climate change. Given the neoliberal order, it seems extremely unlikely that the leading economies will be swayed by Giddens's argument about political and social instability in developing countries. Quite to the contrary, climate change will make them softer targets for economic exploitation. The west tends to follow the dictates of high morality only where abuses of human rights are located in nations of a certain economic clout (Russia, China, Cuba, and North Korea being obvious examples). Otherwise, any amount of political corruption remains acceptable in developing countries that remain 'open for business'.

In *Climate Wars*, Harald Welzer (2012) offers a stark portrayal of the effects climate change is likely to have on international relations in the coming decades. Welzer offers a paradigmatic case of presenting climate change in terms of social catastrophe. Between the alternate futures offered by Klein on the one hand and the United States Department of Defense on the other, he clearly leans towards the latter. His basic argument is that we have so far vastly underestimated the socially catastrophic potential of climate change. He aims his critical remarks chiefly at social and political theorists:

> Climate change will increase the frequency of social disasters, which will bring about temporary or lasting states of society, or social formations, about which nothing is known because too little interest has been taken in the subject up to now. Social and cultural theory is fixated on normality and blind to disasters, but a glance at the cultural history of nature is enough to convince us that it must bring climate change into its purview. Present-day social changes – from the climate war in Darfur to the Inuits' loss of habitat – highlight the startling immateriality of social and cultural theory, and it is high time that it modernised itself and found a way back from the world of discourse and systems to the strategies through which social beings try to control their fate. (Welzer 2012: 26)

The polemic against trends in academic social theory is not of interest here, but the idea of natural resources as an inevitable focus of social conflict is. Welzer's constant parallels between genocide and what he refers to as 'ecocide' shore up his argument that climate change is bound to precipitate all manner of novel deadly social conflicts. In this context his retelling of Jared Diamond's (2005) account of the self-destruction of the Easter Islands' population serves as a parable for the coming climate catastrophe. After reaching a productive peak around 1500 the population of the islands declined over the following two centuries until observed by western explorers towards the end of the eighteenth century in a pitifully weakened state. The statues for which the islands are chiefly known in this context symbolise the environmental folly of a people who plundered limited natural resources with no regard for the viability of future generations. Extensive deforestation had fatally disturbed the ecology of the island, precipitating a devastating cascade of natural resource depletion, which in turn led to decades of internecine warfare and endemic social misery.

This framing of history implies a natural tendency of human societies to exploit the material environment to the point of self-destruction. In other words, it offers yet another rehash of the neo-Malthusian 'tragedy of the commons' narrative. Climate change is then simply this tendency writ large, threatening the human population at the global level. By implication, we will continue in the fashion of the Easter Islanders, allowing fossil fuel exploitation to precipitate a near constant state of international war and self-destruction. Unlike Naomi Klein, who sees the overturning of capitalism as the way to avert climate catastrophe, for Welzer environmental collapse will definitely discredit the Enlightenment faith (shared by Hegel and Marx) that human self-understanding will allow the evolution of a progressively more peaceful and rational world order. Instead, he argues, the most likely scenario under advanced climate change is a world characterised by constant climate wars:

> As climate effects become more extensive and visible, and as hunger, migration and violence grow in intensity, the pressure to find solutions will be more acute and the space for reflection will

be narrower. The likelihood of irrational and counter-productive strategies will become greater, especially in relation to problems of violence exacerbated by climate change. All the historical evidence makes it highly probable that 'superfluous' people who seem to threaten those already enjoying relative prosperity and security will lose their lives in increasingly large numbers, whether from lack of food and clean water, from frontier wars, or from civil wars and interstate conflicts resulting from changed environmental conditions. This is not a normative statement; it simply corresponds to what has been learned from solutions to perceived problems in the twentieth century. (Welzer 2012: 181)

Welzer's analysis is to be commended for the plausible and thorough way in which he demonstrates the historical role played by natural resource scarcity in human violence and war. For all its cogency, however, his account of social history lacks political specificity. For Welzer, the last 250 years have been characterised by an Enlightenment faith in progress that has been dashed at every turn by relentless violence and realpolitik. However, while the origins of climate change do indeed lie in the Industrial Revolution, it is primarily economic growth since the Second World War that has raised levels of atmospheric carbon to their current, highly dangerous point. For Welzer, it is as though the past two centuries of socialist critique of capitalism were nothing but a historical addendum. Through it all, humanity has simply demonstrated its inveterate tendency towards factional discord and environmental imprudence.

Ultimately, Welzer offers simply another variation of the 'tragedy of the commons' theme so central to the neoliberal social paradigm. His attempt to pass off the social problems arising from climate change as a function of an underlying human nature is uncritical in the extreme. It parallels the neoliberal naturalisation of capitalist society. Welzer's conclusions about the likely militarisation of response to climate change do indeed describe the world as it is. But this is the neoliberal world, which has been made and can be potentially, and painstakingly, unmade. Welzer's variety of climate catastrophism has the merit of showing us how the climate drama will likely play out if neoliberal governance remains in place. There are indeed, as he argues, many signs already that across liberal democracies climate change is being used to rationalise an increasingly authoritarian and repressive form of governance. The recent political construction of 'eco-terrorism' is a case in point. Neoliberalism does not offer, as its founding ideologues insisted, a high road to the 'open society'. Increasingly since the end of the Cold War neoliberal governance has shown itself to be capable of the very kinds of political oppression, surveillance and intolerance that characterised the totalitarianism it was originally designed to prevent.

THE POLITICS OF ENVIRONMENTAL INSECURITY

The age of neoliberal environmentalism is also the age of human security concerns. While early modern theories of the state emphasised the sovereign's principal function of protecting the populace from the dangers of external aggression and internal faction, contemporary politics dwells on threats to human security that inevitably outstrip the protective capacities of the nation state. The United Nations first defined 'human security' in its 1994 *Human Development Report*, thereby recognising two aspects:

> first, safety from such chronic threats as hunger, disease and repression. And second, it means protection from sudden and hurtful disruptions in the patterns of daily life – whether in homes, in jobs or in communities. Such threats can exist at all levels of national income or development. (United Nations Development Programme 1994: 23)

Climate change, entailing as it does systemic environmental instability, is clearly a prime cause of human insecurity. Both dimensions of insecurity identified by the UN definition are in play here. In the first instance, agriculture crucially depends on certain reliable patterns of seasonal weather and crop failure, even in an age of global food distribution, can create immediate severe hardship. Secondly, 'hurtful disruption in the patterns of daily life' is also a manifest consequence of climate change, especially in parts of the world where social and physical infrastructure lacks robustness at the best of times. The way climate change is scientifically constructed makes it difficult to accord direct causation for specific events. According to Welzer, however, this complexity should not prevent us from identifying climate change as a cause of violent conflict likely to increase in significance in coming decades. Accordingly, he is impatient of those social analysts who see climate change as nothing more than one background factor among others:

> Environmental changes due to global warming are treated simply as one variable in the interplay of factors that lead to violent conflicts. But this is trivial in so far as neither individual nor collective violence is ever monocausal, and it is even questionable in principle whether the origin and development of violent processes could be given such an explanation [. . .] Nothing that brings human beings to make a far-reaching decision can be traced back to a single cause. (Welzer 2012: 73)

It is important to recall here the complexity of climate change itself as emphasised by Giddens. Given that the phenomenon of climate change is essentially a scientific construct it is unlikely to be explicitly identified by affected individuals and groups as a cause of their decision-making. But while those caught up in conflict may not directly recognize climate change as a cause of conflict, this should not prevent social analysis from doing so. With this in mind, Welzer labels the conflict in Darfur beginning in 2003 'the first climate war' (Welzer 2012: 61). To identify climate change as a decisive factor in this conflict by no means entails, however, that the long history precipitating social fragility should be ignored. As Welzer remarks: 'There had been conflicts for seventy years or more between Darfur's settled farmers ("Africans") and nomadic herdsmen ("Arabs"), but they have become increasingly severe as a result of soil erosion and greater livestock numbers' (Welzer 2012: 62).

In other words, a 'climate war' does not name a condition of armed conflict caused solely by climate change. It is rather the case that, under certain conditions, climate change is a key causal factor needed to precipitate full-blown war where lower levels of social dispute were already present. Given that climate change cannot be straightforwardly singled out as *the* cause of this or that conflict, it is likely that economically dominant nations will have recourse to subterfuge: either to deny directly that climate change is a key cause of human insecurity or insist that it can only be identified as a cause of conflict in terms of a natural (that is, not a humanly caused, ultimately politically induced) catastrophe. The leading neoliberal states have exhibited striking levels of denial over their historic role in causing climate change. In international negotiations the very suggestion of economic reparations to be paid by the early industrialising nations to those poorer countries most affected by climate change has been treated with contempt. This pattern of denial is hardly likely to change in the more contentious context of armed conflict.

Welzer's (2012) analysis of contemporary armed conflict also underscores the phenomenon of forced migration due, at least in part, to climate change. He remarks:

> The very category of 'climate refugee', so hazy in international law, makes clear that the decision to flee may result from war, massacre, extreme weather, rising sea levels or loss of a subsistence base; several of these typically come together when people decide to seek salvation elsewhere. (Welzer 2012: 73)

The neoliberal order evinces ambivalence, amounting to a full-blown contradiction, in relation to cross-border migration. As Foucault points out in his lectures, in many ways the migrant can be viewed as the archetypal neoliberal entrepreneur. Social mobility is readily manifested by geographic mobility and, within the borders of an affluent nation state, the idea that the worker comes to the job rather than the job to the worker constitutes a key neoliberal norm. Foucault (2008) captures the neoliberal analysis of migration in the following way:

> Migration is an investment; the migrant is an investor. He is an entrepreneur of himself who incurs expenses by investing to obtain some kind of improvement. The mobility of a population and its ability to make choices of mobility as investment choices for improving income enable the phenomena of migration to be brought back into economic analysis . . . in terms of individual enterprise, of enterprise of oneself with investments and incomes. (Foucault 2008: 230)

Between Welzer and Foucault we can identify two obvious tensions: first, between the forced migration connected to climate change and the voluntary migration valorised by neoliberalism; and, second, between cross-border migration as opposed to internal or domestic migration. As is often remarked, advanced capitalism may have given rise to a world where commodities circulate around the globe ever more freely but this is only selectively the case for human capital. In many ways freer global circulation of goods necessitates restricted freedom of movement for global labour, for the obvious reason that profit is generated by the coexistence of poor productive labour and relatively wealthy consumers. In a borderless neoliberal world hundreds of millions of people from developing economies would doubtless make their way to more affluent developed nations. This would precipitate almost immediately a critical shortage of labour power in the poorer nations, forcing one of two solutions. Either production would have to follow the fleeing labour or wages would have to rise considerably for those who remain within the poorer nations, now depleted of their 'reserve army' of labour. In either case the cost of production would rise significantly.

Thus it is clear that neoliberal capitalism must generally oppose the free movement of labour. The real world policy solution to neoliberalism's ambivalence on this issue is legal migration generally restricted to those possessing labour skills of value to the host nation. Countries such as Canada and Australia, for example, have instituted a points system whereby potential economic migrants are rated according to identified needs within the workforce of the receiving nation. This seems an elegant and practical solution, allowing controlled migration of a steady stream of high-value human capital. The obvious problem when we connect this back to climate change is that we can no longer plausibly talk of voluntary migration when it comes to the kind of hardship brought about by climate instability. It is much more a question of push rather than pull.

Economic migration when thought of as largely voluntary can be seamlessly integrated into the neoliberal social imaginary of benign mutual competition. Climate change, by contrast, does not generate economic migrants but rather refugees seeking a place of safety from environmental insecurity. While it can always be argued on neoliberal economic terms that it would be better on balance for that individual worker to remain in her native developing country, it is harder to argue that a person could or should rationally choose to stay in a place exposed to aggravated climate change. Only the harshest of Malthusian arguments would insist on the logic of such human sacrifice. As many analyses in environmental ethics would argue, on the contrary, granting extensive asylum to those at the sharp end of climate change would represent a valuable act of international justice given developed nations' key role in causing climate change in the first place (see Biermann and Boas 2010).

The neoliberal reality is, however, quite otherwise. Just as climate change is already being configured as a national security issue by the most powerful neoliberal states, so too is the phenomenon of climate migration being rapidly brought into the fold of the militarised security agenda. The ongoing 'migration crisis', involving hundreds of thousands displaced by conflict in the Middle East and North Africa and seeking entry to the European Union, is an obvious case in point. A large proportion of those wishing to enter the European Union Schengen Area (within which freedom of movement across twenty-six countries of mainland Europe is allowed without passport checks) are fleeing the protracted civil war in Syria. Millions of Syrians are already living in neighbouring countries such as Turkey and Lebanon, yet the much more affluent countries of the European Union are largely inclined to deny entry to significant numbers of refugees.

Those countries most opposed, such as the United Kingdom, argue that it would be better if everyone's case for asylum were processed individually in centres set up outside the European Union (see Wintour 2015). Welzer traces the origin of this preferred solution back to a 2003 paper entitled *New Vision for Refugees* and produced by the Blair government in the United Kingdom. The linchpin of the report's recommendations was the creation of 'transit processing centres' (TPCs) beyond EU borders that would be dedicated to processing asylum applications efficiently and sending failed applicants back to their country of origin. While the report initially met with vigorous opposition from certain EU member states the basic idea of holding and processing potential asylees beyond European borders has gradually become a key element of EU migration policy. Keeping would-be migrants in their country of origin brings with it the additional benefit of underscoring the popular first world message: your economic disadvantage is not our problem.

As Giddens (2011) argues, the kind of social instability produced by climate change should be considered a problem by wealthy nations for purely self-interested reasons. But the actual tendency of political decision-making points in the opposite direction. Whether motivated by direct economic exigency or in order to escape a worsening climate, migrants from poorer countries are generally seen as something to keep outside the gates of the affluent world. With the present reality of tens of thousands of migrants drowning each year in the Mediterranean in a bid to reach EU territory, the preferred neoliberal response is to blame human traffickers for exploiting desperate individuals fleeing conflict. This is a clear case of attacking the consequences rather than the causes of migration. Were there no violence and hardship to escape presumably the human trafficking infrastructure would disappear almost immediately. The neoliberal logic obfuscates the issue by conflating the voluntary and involuntary aspects of cross-border migration. This contradiction of neoliberalism often comes to the fore when human rights obligations to 'genuine' asylum seekers are played off against the claims of those identified as economic migrants. If neoliberalism remained true to its own entrepreneurial social vision it would actually declare international economic migration as the 'natural' global condition and institute a world of open borders.

BEYOND NEOLIBERAL ENVIRONMENTALISM

The reality of uneven development requires that we remain in a world of at best partially porous borders. In a self-contradictory manner, neoliberal governance has simultaneously constructed the crisis of an ageing population in most affluent countries and a crisis of overpopulation and youth unemployment in the rest of the world. While the former is used to justify austerity politics by pointing to an increasing proportion of 'unproductive' human capital, chronic unemployment in developing countries is ascribed to a relative lack of economic development. On both sides of the developed and developing world divide human capital is deemed to be underperforming. While the ultimate social value of high-performing neoliberal labour remains elusive one thing remains clear: the uneven development necessarily produced by neoliberal capitalism will be exacerbated by climate change. More strident warnings of global crisis will be issued as a consequence. We will, no doubt, hear more about how groups dedicated to fomenting terror are exploiting climate change as a source of social instability. Climate refugees and migrants from developing nations will be increasingly presented as a threat to the limited natural resources of more affluent nations.

One thing that we are unlikely to hear is that climate change is a symptom of neoliberal governance. Instead, focus will remain on the social effects rather than the political-economic causes of climate change. While institutions advocating universal human rights will continue to plead for fairer treatment of those hardest hit by the consequences of climate change, proponents of the SD agenda will insist that more neoliberal development is the only viable path for developing nations. Perhaps, as Klein (2014: 256–90) suggests, failure to diffuse this tension will lead to a spree of environmental entrepreneurialism in the form of unregulated 'geoengineering'. Lacking a credible political alternative to the neoliberal order many may even welcome the prospect of such technologically driven, democratically unaccountable 'solutions' to climate change.

Engaging with climate change makes futurology an enticing pastime. It is not my intention in this book to add to the growing genre of environmental catastrophe narratives. Arguing that this tendency simply aids neoliberal governance is a basic contention of the current analysis. Instead, the purpose of this chapter has been to identify the sustainable development paradigm as the framework of environmental neoliberalism. Recalling Foucault's assessment, it was pointed out that neoliberalism has a peculiar interest in the human environment as a key determinant of social-economic activity. The point here was to unmask the apparent political neutrality and good sense of SD and demonstrate that it represents, in fact, a neoliberal articulation of environmental concern.

In making this argument I am, of course, aware that there are countless environmental initiatives and activists who identify with the sustainability agenda but distance themselves from neoliberal capitalism. For them, no doubt, my critique involves very broad brushstrokes and drastically underplays the diversity of approach gathered under the environmental sustainability umbrella. My response to this is to remind those who are attached to the SD paradigm that there is no environmentalism worthy of the name that does not entail radical social critique. As mentioned at the beginning of this chapter, modern environmentalism began life as such a critique but has, over the last three decades, been co-opted by leading corporations and rendered a banal, technocratic element of political common sense. This process of levelling has occurred even as the dramatic consequences of climate change had become better understood.

The question engaged throughout this book is: what does climate change really amount to as a social-political problem? Through the lens of neoliberalism climate change represents a technical challenge to entrepreneurial innovation; it is a puzzle to be solved by the great corporate leaders in concert with well-meaning scientists and market-friendly politicians. This appreciation of climate change does not call for a popular environmental movement but rather for a cabal of unaccountable experts whose political legitimacy is based on the paternalistic logic 'we know best'. Genuinely democratic politics cannot but stand opposed to this paternalism.

Neoliberal SD thinking entails the eclipse of the democratic project. Collectively shaping the material environment in which everyday life goes on is the key claim of radical democracy. The problem of climate change, identified and articulated by highly specialised scientists and then acted upon (or not) by politicians in consultation with corporate leaders, leaves democratic society with nothing to do beyond idly commenting on events. Originally, modern environmentalism was an organised civil society protest movement against the ecological consequences of unbridled capitalist development. Neoliberalism's demolition of the living history of the workers' movement has largely succeeded in cutting the demos off from its radical political past. The real challenge of climate change is to take up once against this critical condition and reshape social organisation in genuinely democratic ways.

The basic conclusions of this chapter are therefore the following:

- The sustainable development paradigm is at base neoliberal environmentalism
- The rise of climate change should be seen as a symptom of neoliberal governance, that is, as an aspect of neoliberal social pathology
- Environmental activists responding to climate change should abandon the SD paradigm (in all its forms) in an effort to undermine neoliberal governance
- Climate change should be grasped as a political problem that calls for state agency freed from the corporate-led neoliberal model
- Under neoliberal governance instances of climate war and climate migration will proliferate but be subordinated to the logic of the neoliberal security agenda
- Such logic ensures that there will be no genuine international cooperation to counteract human suffering caused by climate change.

The first three chapters of this book have laid out the problem of climate change in the age of neoliberal governance. From an initial analysis of the inherently political status of nature in Chapter 1, we considered the uneven development inherent in the capitalist production of nature in Chapter 2, before turning to a more explicit account of the nature of neoliberalism in this third chapter. If the first three chapters offer a kind of deconstruction of neoliberal environmentalism, the final two attempt to outline a counter-neoliberal environmental politics.

Here I rely on recent attempts by David Harvey and other allied thinkers who seek to recuperate Henri Lefebvre's idea of 'the right to the city'. I have already frequently anticipated this move when referring to the environmental commons as something neoliberal governance is inherently incapable of acknowledging and safeguarding. Neoliberalism actively pursues the 'tragedy of the commons' when it comes to management of any shared environment. In the next two chapters I argue, by contrast, that there is nothing inevitable about this tragedy. In fact, social history shows that truly 'sustainable' societies are those that have been characterised by the cooperative protection of shared sites and customs. As such, we can speak of an older 'comedy of the commons' lying beyond the horizon of the neoliberal order.

CHAPTER 4

Environmental politics and place

REFRAMING SOCIAL SUSTAINABILITY

The concept of sustainability has been at the centre of environmental policy, science, and more general debate for the last three decades. While arriving at a coherent definition has remained elusive, there is general agreement that sustainability ideas and practices are the best way to tackle the most pressing environmental challenges. Part of the attraction of sustainability is its apparent neutrality: no particular political or social vision is attached to it, allowing thereby various stakeholders to align different, often irreconcilable ideals of a desirable future. This neutrality, as I argue throughout this book, is illusory. In fact, the idea of sustainable development (SD) is specifically configured to provide powerful ideological justification for neoliberal economics and politics. The dominance of the SD paradigm in public discourse and policy accounts for the fact that, despite constant warnings that we are approaching environmental limits on a bewildering number of fronts, there is no coherent vision of a credible alternative, that is, counter-neoliberal organisation of society.

The fact that the World Trade Organization (2016) has, for some time now, proclaimed sustainability as one of its core principles is a clear sign that the sustainability agenda is in reality an ideological justification for neoliberal economics. Despite the fact that global equity has been a formal part of international sustainability policy since the Brundtland Report of 1987, in reality the same patterns of global economic exploitation have prevailed and significantly accelerated in the intervening decades. While capital has chased cheap labour across the globe, the social and environmental conditions of workers have been severely impacted (see Harvey 2005: 87–119). The largely enforced urbanisation of developing nations is touted as a fast track to western style wealth but, as Mike Davis (2005) made clear, this has in actual fact produced a 'planet of slums'. Neoliberals and progressive liberals alike have praised the mechanisms of 'microfinance' for bringing the rural poor (especially women) of developing nations into the exchange economy. But this can also be seen as a further effective way of tapping into the reserve army of cheap labour always sought out by capital (see Davutoğlu 2013).

In short, there can be little doubt that the paradigm of sustainable development has offered significant ideological support to the hegemonic global consensus on how all manner of resources (natural and human) are to be harnessed and deployed. But this involves at its core a very specific political framing of nature, including the nature of human potential. This framing of nature has effectively effaced or appropriated more radical social visions of earlier environmental movements, thereby integrating them into the common sense of neoliberal governance. Accordingly, in affluent liberal democracies environmental issues are now predominantly framed as consumer responsibility problems. If only the populace could be properly informed of their deficient actions and induced into more benign patterns of consumption, so the argument goes, the efficiency savings accrued would go a long way to meeting some key environmental challenges. For developing economies, similarly, the overarching goal is to arrive at a western-style consumer society by means of educational infrastructure (especially for women) that will raise the opportunity costs of having large families. The hubris and condescension of this neoliberal development stance lie barely below the surface (see Harvey 1996: 146–9).

My central contention here is that, in opposition to this hegemonic understanding, what is fundamentally needed to meet the most pressing environmental challenges is the construction of a cogent counter-neoliberal social imaginary. Central to this task is an effort to rethink the nature of what is bureaucratically referred to as human resources. Within the sustainability paradigm, it is the so-called third pillar of social sustainability that should answer to the questions of work and labour. But the same neoliberal predetermination that is evident with regard to the other two pillars of economic and environmental sustainability inevitably characterises the SD concept of the social. The analysis offered here seeks to overturn this understanding of social sustainability.

The first thing to note is that, properly conceived, social sustainability crucially involves consideration of the interconnections between human activity and place. In other words, it is a question of the interface between work and place. A salient feature of much valuable environmental praxis is its use of concrete places as sites of resistance against the socially corrosive and regressive forces of neoliberal globalisation. The politics of place is useful for combatting the false neutrality of officially sanctioned sustainable development. As Neil Smith (2008) so eloquently clarified, capitalism cannot deliver on its promise to raise living standards simultaneously across the globe, but can only shuffle the pack in its regime of uneven development.

One of the ideological consequences of the political dominance of neoliberalism is a virtual eclipse within the popular imagination of the rich history of alternate models of work. The neoliberal disciplining meted out to workers in developed economies in the process of deindustrialisation has been very effective in extinguishing any sense of international working-class solidarity. The precariousness of employment in even the most affluent societies serves to keep the average worker trapped in a cycle of individualised self-preservation. A relentless ideological war waged against organised labour has further undermined the efficacy of the working class within mainstream politics across affluent liberal democracies.

Against this backdrop, appealing to natural resources (such as fossil fuel reserves) as a global commons becomes politically ludicrous given the realpolitik of geopolitical rivalries. In the context of advanced globalisation, the population of any current nation state, whatever its polity, can hardly be regarded as having credible collective control over energy reserves. The citizenry of the great consumer nations, such as the United States, do not possess any greater degree of popular political control despite the repeated admonitions to realise consumer power. As the current race to exploit fossil fuel reserves in the Arctic amply illustrates, it will not be 'we, the people' who decide what stays in the ground and what does not.

In this context it is striking that the same metaphors derived from the Malthusian appreciation of the human relationship to natural resources suffuse the dominant discourse relating to work. Rather than work being viewed as a potential source of voluntary social experimentation and cooperation, workers are constantly cajoled to prepare themselves for the harsh conditions of market competition. The permanent spectre of potential redundancy is one of the many manifestations of alienation that have long since become given features of the neoliberal reality of work. Marx cut through this ideological construction by demonstrating how the real process in question is one of extracting 'surplus labour' from workers, namely that portion of the value of work that is not returned to the worker in the form of pay or other benefits (see Sitton 2010: 109–10).

The normalisation of worker precariousness in societies where wealth production is high (most recently illustrated by austerity politics in Europe and elsewhere) is actually a feat that requires constant effort. One dimension of that effort is reinforcing the appearance that those who do best out of the capitalist system possess rare natural talents whose scarcity justifies the outsized rewards they reap from work. Average workers, by contrast, are required to internalise a self-image of worthlessness, according to which they should consider themselves grateful or lucky for any employment that keeps them above basic subsistence levels. When workers demand increased wages, more respect, or better working conditions, employers and their ideologues are always ready with the riposte: we would love to grant all these things, but it will hurt our efficiency and, ultimately, you. While unionised workforces that campaign for higher wages are easily pilloried by corporate owned media as greedy, self-serving workers making things harder for others, the more profound injury sustained by workers in general is a permanent sense of social redundancy.

Thus, there is a profound homology between the discourse of natural resource scarcity and employment scarcity. Given the aura of inevitability on both sides it takes a deliberate effort of thought to realise that work scarcity is in all cases a manufactured phenomenon. If one conceives of a market as simply a circumscribed social organisation of goods and services, then the only genuine restriction one need recognise is providing for all basic needs. If, further, one imagines the place in question as strictly delimited, then problems of overexploitation are almost inconceivable. Given the potential productivity of all resources (natural and human), the only concrete question that could arise would relate to desirable actions beyond those necessary to meet basic needs. Marx did not offer answers to this question of non-necessary production (he was deliberately vague about the post-revolutionary social condition), for the simple reason that it would make no sense to do so. His point was that the capitalist mode of production will never create material conditions where humanity can even pose the question of optimal human flourishing in earnest. The 'struggle for existence' will instead continue to preoccupy worker and capitalist alike.

THE PROBLEM OF WORK

A famous slogan of the Paris mass protests of 1968 was 'never work'. These words were reputedly written by Guy Debord in 1953. In 1957, Debord and others close to him formed the political-artistic collective the Situationist International (SI). In its early years the group designed alternative, collectively experimental ways of experiencing the urban environment. Taking its cue from an earlier artistic collective, the French surrealists led by André Breton, the SI sought to overturn in practical ways the utilitarian logic of bureaucratic urban planning. In these efforts the group was drawn to the conception of 'homo ludens' (the playful human) and the idea that the efficiency model of modern

work needed to be displaced by a society focused on playful collective action. In this the SI also represented a critique of mainstream Marxism's focus on revolutionising the social control
and benefits of industrialised labour. The injunction to avoid
work gave voice to a sense that modern work could no longer be
salvaged and transformed into something positive for industrial
consumer society as a whole.

Around the same time, the thinking of the important Frankfurt
School member Herbert Marcuse was tending in a similar direction. Notes for a lecture course delivered in Paris in 1958–9 sketch
out some of the basic ideas that were later developed in Marcuse's
widely read book *One-Dimensional Man* (1991 [1964]). In the
notes Marcuse highlights the essentially repressive character of
modern work:

> Civilization is man's subjugation to work. In this process, the
> human organism ceases to exist as an instrument of satisfaction
> and instead becomes an instrument of work and renunciation
> [. . .] Society therefore must turn the instincts away from their
> immediate goal and subjugate them to the 'reality principle',
> which is the very principle of repression. (Bronner and Kellner
> 1989: 125)

The important point expressed here is the idea that modern work
is not intrinsically connected to the satisfaction of the worker.
Let us recall here Foucault's (2008: 226) analysis of the neoliberal rationality of acts of consumption, as 'work' productive of
individual satisfaction. This hints at the way in which neoliberalism would succeed in placating widespread social dissatisfaction
with work.

In more general terms, the core task of the Frankfurt School
could be characterised as bringing about a critical theory of
modern society and culture on broadly conceived Marxist foundations. The rejection of utopian socialism was a function of
Marx's disinclination to project alternative, post-revolutionary
forms of work, thereby creating a gap that Marcuse and others
were able to fill in the context of the 1950s and 1960s. With
the waning of grassroots radicalism and transition to the global
neoliberal conjuncture following the Oil Crisis of 1973, efforts
to envisage and realise more radically democratic frameworks of

work have been largely eclipsed within western liberal democracies. This is unsurprising given the underlying effects of neoliberal policies: dramatic fluctuations in mass unemployment; the increased precariousness of work tenure and conditions due to the dismantling of union protections; and the extremely successful rhetoric of raw economic growth as measured by gross domestic product. After all, why worry about the nature of work when keeping hold of any kind of job is a daily worry; and why worry about falling wage levels when consumer goods from overseas are getting cheaper all the time?

While the recent actions of the Occupy movement in the United States and mass protests against austerity policies across the European Union have received mass media attention, the fundamental problem of the nature of work under neoliberalism has not so far loomed large. Casting the struggle as everyone else against the wealthiest 1 per cent amounts to a case for better distribution, not for revolutionary changes in how wealth is produced in the first place. This ultimately misses the point of challenging neoliberalism, which has been enormously successful in blunting the edge of effective social critique through its unparalleled capacity to create wealth. Just as importantly, improved wealth distribution would do little or nothing to tackle the source of the environmental problems we face. The reality of climate change calls for social action far more radical that a traditional social democratic fairer distribution of the spoils of capitalist development.

A far more promising path takes us back to the original focus of radical socialist critique, namely the centrality of private property to capitalist production. The environmental systems affected by climate change represent the ultimate commons. Social and political organisation must be seen as a matter of common interests, not only of the human species, but rather of all forms of life. Such is the generally recognised goal of the environmental movement. The underlying problem with neoliberal governance, seen from an environmental perspective, is not the unfairness of its distributive mechanisms, but how it produces wealth in the first place. Even if a shift from carbon-based energy sources is achieved in the future, an intact neoliberalism would continue to produce wealth without eliminating the environmentally harmful externalities that are intrinsic to capitalist development. All efforts, practical and theoretical, should therefore be focused on the task of reclaiming natural systems as a commons. To get an initial sense of what this task involves it is useful to turn our attention, initially, to the urban environment. For it is here that the pathologies of neoliberal development are most prominently on view and most readily contested.

THE URBAN PROBLEM

To more effectively tie together work and place I will follow the lead of Harvey (2012) by reflecting on the urban condition as exemplary both with regard to the concrete conditions of contemporary neoliberal capitalism but also as a context for conceiving progressive alternatives. Using the idea of the 'right to the city' as a currently influential conceptual frame for grassroots urban democracy, I argue that this conception can also allow us to conceive of work in ways quite opposed to the current neoliberal reality. Following this line of inquiry leads to the following conclusion: the orthodox concept of social sustainability needs to be replaced by a model of worker-controlled production. This is the straightforward application of the right to the city idea to the question of free labour.

By contrast, the actually existing form of sustainability practice is characterised by top-down, expert decision-making and implementation. In other words, it endorses technocratic rather than democratic control over material production. Social sustainability, nested as it is within the broader paradigm of SD, cannot lead to a condition of free labour as it is predicated on a putative condition of material scarcity that can only be resolved through socially restrictive and politically repressive technocracy. SD always looks ahead to a stabilised condition of global wealth production that it cannot possibly deliver. For all its insistence on being the only reality worth attending to, the neoliberal growth paradigm is at base a utopian social scheme far less credible than anything dreamt up by two centuries of socialist thought.

The French Marxist sociologist Henri Lefebvre originally formulated the idea of the right to the city in the momentous year of 1968, which saw massive social upheavals involving workers and students across Europe and beyond. Lefebvre was closely associated with the SI during the initial years of the group's activity between 1957 and 1960 and the impact of the Situationist critique of bureaucratic urbanism is readily apparent in his writings on cities. What makes Lefebvre's idea more than a historical curiosity is the fact that it has been recently taken up by a number of contemporary theorists, most notably by David Harvey. In his recent book *Rebel Cities* Harvey (2012) explains that asserting a right to the city means 'to claim some kind of shaping power over the process of urbanisation, over the ways in which our cities are made and remade, and to do so in a fundamental and radical way' (Harvey 2012: 5).

In work going back to the early 1970s, Harvey has charted the democratic deficit that has characterised modern urban planning since its inauguration in Paris under the auspices of Baron Haussmann in the 1850s. Based on a detailed analysis of Marx's account of political economy originally set out in *The Limits to Capital*, Harvey (2006 [1982]) excels at demonstrating how bureaucratically controlled urbanisation has one basic driver: the desire to extract further profit through ground rent. This has many directly negative social consequences, the most obvious of which is the displacement of the working class from more desirable and better-served urban areas. In *Spaces of Hope*, for example, Harvey (2000: 133–56) turned his attention to the recent history of Baltimore to show how urban regeneration has been a major cause of exacerbated material inequality in that city.

While Harvey readily recognises the limitations of appeals to rights (he often cites Marx's dictum that disputes between contending rights tend to be settled by superior force), he still believes such appeals are a powerful way to address social injustice. The point about the right to the city is that its satisfaction requires demonstrable popular control over the material context of urban life. It rests on the decisive insight that popular, democratic political agency is not credible in the absence of control over the shaping of the built environment. Harvey also argues that increasingly sophisticated surveillance and oppression in our cities in part stem from an intensified official intolerance of the democratic occupation of public space. As the recent Occupy movement once again made clear, the public's practical assertion of a right to the city directly challenges the security agenda that states typically use to justify severe restrictions on freedom of assembly. More banal efficiency objections are also common, as when the city of Portland, Oregon, justified its forced eviction of Occupy activists on the grounds that they were burdening the city's parks and recreation budget at a time of fiscal austerity. Local media outlets, ever the champions of neoliberal common sense, were successful at projecting the indignation of the silent majority of Portland residents.

Thus the contemporary city across liberal democracies exhibits contradictory characteristics. A protracted and mostly subterranean conflict is taking place between the neoliberal forces of urban redevelopment and a heterogeneous contingent of social opposition. 'Gentrification', a phenomenon characterising many cities across wealthy liberal democracies, is one of the effects of neoliberal development that is more overtly discussed and, increasingly, opposed. As real estate values rise, displacement of poorer residents is inevitable, thereby raising the question what right, if any, a given population has to remain in a specific place. The language of rights, as Marx was apt to point out, is the language of the bourgeois revolution that brought about the French Republic. Rights are traditionally accorded to the individual, such as the right to free speech, freedom of conscience, etc. The right to the city, by contrast, is both an intrinsically collective right and something more inherently political rather than legal or moral in nature. The right to the city in essence calls for a radicalised form of urban democracy, such that those who live within the city get to determine the kind of city they reside in.

This stands opposed to the current reality of the bureaucratically determined city, whereby elected or appointed municipal actors engage in various forms of urban planning and facilitate business interests. The economic imperatives that direct national governments are just as pressing if not more so at the municipal level. In *Constructing Community* (Elliott 2010),

I analysed how municipal public outreach typically offers little that plausibly counts as genuinely citizen-led decision-making. Instead, when a metropolitan bureau of planning 'reaches out' to an affected public, the options have typically been predetermined in the name of expert knowledge or corporate interest. The idea of the right to the city stakes a claim that directly challenges this pastiche of public participation. The right in question is one of radically democratic urban self-determination. Just as, according to Smith (2008), we should look on nature as something produced rather than given, so too the nature of this or that city is a result of dominant forces of production.

FROM URBAN SUSTAINABILITY TO THE RIGHT TO THE CITY

The idea of the right to the city provides a useful antidote to the concept of social sustainability principally because the environmental goods in question are a pre-eminent instance of a commons. The ubiquitous 'tragedy of the commons' notion – which insists that resources held in common will inevitably be exploited to the point of exhaustion – follows the familiar pattern of neoliberal thought. Accordingly, the only way to protect the environment is to leave it to the dynamism of market mechanisms to arrive at the true price of resource exploitation. Thus, the dominant solution to climate change pursued in recent decades has been the creation of carbon trading markets.

The problem with this is obvious: the corporate players involved in fossil fuel extraction are never going to price carbon high enough to curtail significantly or phase out the use of this source of energy. This would be tantamount to expecting them to direct their efforts towards diminished profit making. But this, the IPCC insists, is precisely what needs to happen. Affluent liberal democracies pride themselves on the superior standards of environmental health they have achieved since the 1970s. But the benefits of better air quality, such as they are, hardly weigh in the balance when compared to the predicted negative impacts of climate change. Given the obvious inadequacy of neoliberal market mechanisms in dealing with climate change, the question must be asked whether the nation state is doomed to enact the tragedy of the commons.

The idea of the right to the city allows us to start out on a path that directs our thinking away from the Malthusian assumptions of the tragedy of the commons. In the next chapter we will sketch out an antidote to this, a veritable 'comedy of the commons'. For now, let us return to our point of departure in this chapter and ask: How does to the right to the city lend direction to a reconfiguration of work in the spirit of genuine social sustainability? Lefebvre's original conception of a right to the city regards the urban environment as the premier site of capitalist class conflict. As Harvey has similarly contended in his work, a key way in which the neoliberal disciplining of society works is through the management and control of the urban environment for maximised profit-making. Accordingly, the 'urban renaissance' that western cities have enjoyed since the 1980s in reality arises out of the efforts of financial and corporate interests to realise profit through intensified construction and extraction of ground rent.

In reconstructing the history of neoliberal urbanism, Harvey (2012: 11–14) highlights the exemplary nature of the fiscal crisis of New York in the 1970s insofar as it led to the sort of realignment of urban politics and economics that has since come to dominate the regeneration of major cities across the United States and elsewhere. The right to the city is essentially about genuinely democratic determination of the urban environment. By contrast, urban sustainability as it is currently practised is concerned with consolidating bureaucratic control against the claims of such popular control. The right to the city calls for democratic accountability as opposed to the currently hegemonic system of bureaucratic accountability. In effect, this returns the socialist and anarchist critiques of industrialised society to their origins. It also invokes the class struggle in its original formulation as a radical conflict between workers and capitalists. In this context, it is clear that the mediating function of labour unions has been severely curtailed across western nations in the last three decades. As of 2013 union membership in the United States stood at a little over 11%, down from around 20% in 1983 (US Department of Labor 2014). Grasping what this means in material terms for American labour requires a better understanding of how capitalist production is predicated on corporate and bureaucratic control of land and the built environment.

In *The Limits to Capital*, Harvey (2006) drew attention to the pivotal significance of such control in terms of the capitalist extraction of ground rent. He quotes Marx from *Theories of Surplus Value*:

> If the land were . . . at everyone's free disposal, then a principal element for the formulation of capital would be missing . . . The 'producing' of someone else's unpaid labour would thus become impossible and this would put an end to capitalist production altogether. (Harvey 2006: 359)

This serves as a point of departure for Harvey's own analysis of the conditions of modern urban development: 'The theory of ground-rent resolves the problem of how land, which is not a product of human labour, can have a price and exchange as commodity' (Harvey 2006: 367). Not only is capitalist control of land through rent extraction a principal instance of the alienation of workers from nature, it also leads to the extremely volatile but crucial formation of what Harvey terms 'fictitious capital'. Surplus value generated through legal control of land is not fictitious in the sense of subjectively imaginary or unreal but rather insofar as it stems from 'a claim upon anticipated future revenues, a claim upon the future fruits of labour' (Harvey 2006: 367). As Harvey (2011) has reiterated more recently: 'The power of land and resource owners has been much underestimated . . . This area of activity accounts for as much as 40 per cent of economic activity in many of the advanced capitalist countries' (Harvey 2011: 182).

The point of enumerating these observations is to make clear how the neoliberal urban environment is fundamentally characterised in two ways: first, as an intensified space of capitalist surplus production; and, secondly, as a domain of starkly undemocratic control of space and place. This brings us back directly to the right to the city. As Don Mitchell (2003) points out in his analysis of urban politics, central to Lefebvre's original presentation of the idea is the 'normative argument that the

city is an *oeuvre* – a work in which all its citizens participate'
(Mitchell 2003: 17). As Mitchell also notes, modern urban
planning's appeals to participation largely do not constitute
credible contexts of residents' active production but rather
their passive consumption of space. He quotes Guy Debord
from *The Society of the Spectacle*: 'The proletarian revolution
is that *critique of human geography* whereby individuals and
communities must construct places and events commensurate
with the appropriation not just of their labour, but of their
total history' (Mitchell 2003: 19). This critical observation
goes back to the founding of the SI in 1957 and is explicitly set
out by Debord in his 'Report on the Construction of Situations'
from the same year. Here an experiential appreciation of the
urban environment is proposed under the guise of a prospec-
tive 'psychogeography'. The goal of such an approach to the
built environment, the young Debord relates, is 'to reduce the
empty moments of life as much as possible' (Knabb 1995: 23).
He goes on: 'One could thus speak of our action as an enterprise of quantitatively increasing human life', which 'implies a
qualitative increase' (Knabb 1995: 23–4).

It is easy to dismiss the Situationist position as a mere product of youthful exuberance and naivety. But it has proven to be
both an explicit and implicit source of inspiration for popular
urban political movements up to and including 2011's Occupy
Wall Street. What earlier movements such as Reclaim the Streets
had already made clear is that it is not enough to contest capitalist domination of the urban environment: a positive alternative must also be demonstrated. The ideologues of neoliberalism
such as Hayek and Friedman were very effective in shaping a
ruling conception of liberty in terms of the absence of external
constraints, primarily in relation to state control of markets.
This allowed the removal of restrictions on the transnational
movement of capital, corporate ownership, and the ability to
seek out the most readily exploitable labour force to pass for the
inevitable processes of modern freedom. As long as consumer
goods could be made available to residents of wealthier countries at progressively lower prices, this could count as making
evident the fact that capitalism delivered a higher standard of

living. But if this consumer paradise was already recognised by the likes of Marcuse and Lefebvre as a travesty of genuine freedom under the significantly better economic circumstances of the 1960s, then this case must be made all the more emphatically in current conditions of enforced neoliberal austerity. This calls for a thoroughgoing re-evaluation of the connection between genuine quality of life and the conditions of work.

ARENDT ON THE MARXIST CONCEPTION OF LABOUR

This question is central to Hannah Arendt's (1998 [1958]) major work *The Human Condition*. What marks out Arendt's thinking is her principled critique of the Marxist conception of labour. Given the underlying Marxist character of the critique of urban politics advanced thus far, the nature of Arendt's critique of Marx can offer important clues about the potential weaknesses of attempting to reinterpret social sustainability by means of the right to the city conception. While Arendt's critique can, I believe, ultimately be adequately answered through further reflection on the socialist ideal of worker-controlled production, her objections help us to test the credibility of the right to the city idea. At the very least, Arendt's account of work reminds us that the appeal to intergenerational justice, which lies at the heart of the sustainability paradigm, amounts to a genuine concern that must be incorporated within any counter-neoliberal political ecology.

In an early section of *The Human Condition* Arendt characterises the private sphere in terms of action aiming at the maintenance and survival of human life. The public sphere, by contrast, is constituted by the properly political concern for the display of excellence and virtue. While the former involves actions directed towards the constantly present needs of biological survival (in ancient Greek, human existence as *zoe* or 'mere life'), the latter aims at the 'immortality' of achievements that outlast the individual and the present generation (in Greek, human existence as *bios* or 'political life'). Following Aristotle's account of democratic politics as a society of equals, Arendt describes the ancient city-state in terms of an intense struggle between individuals for the recognition of excellence:

> To belong to the few 'equals' (*homoioi*) meant to be permitted to live among one's peers; but the public realm itself, the *polis*, was permeated by a fiercely agonal spirit, where everybody had constantly to distinguish himself from all others, to show through unique deeds or achievements that he was best of all (*aien aristeuein*). The public realm, in other words, was reserved for individuality; it was the only place where men could show who they really and inexchangeably were. (Arendt 1998: 41)

The public sphere as the forum in which equals vie for socially recognised distinction is also the domain of the 'common'. While Arendt thus makes commonality a function of individual competition, she explicitly rejects the notion that oppression or force is present. Accordingly, she defines the political domain as 'the sphere of freedom' as opposed to the private household 'born of necessity' (Arendt 1998: 30). This already makes clear how Arendt's conception of politics is starkly opposed to Marx's notion of class antagonism relating to the means of production.

While Arendt's appeal to an 'agonistic' model of public life superficially resembles neoliberal conceptions that celebrate economic competitiveness, in fact she accounts for political action as a desire to overcome the limitations of individual existence. As she remarks: 'For the *polis* was for the Greeks, as the *res publica* was for the Romans, first of all their guarantee against the futility of individual life, the space protected against this futility and reserved for the relative permanence, if not immortality, of mortals' (Arendt 1998: 56). Arendt's main point against Marx is that his conception of 'labour power' (*Arbeitskraft*) allows him to align labour and production, thereby conflating what Arendt takes to be a crucial distinction between labour and work. Marx's desired social revolution would result, Arendt contends, in a situation where 'the distinction between labour and work would have completely disappeared' (Arendt 1998: 89). The consequence of this is that 'all work would have become labour because all things would be understood, not in their worldly, objective quality, but as results of living labour power and functions of the life process' (Arendt 1998: 89).

Arendt draws here on the German connotations of 'work' in such terms as *Kunstwerk* (artwork) to insist that it denotes designed production of useful and beautiful objects. Marx, by contrast, from his earliest writings on pursues an anti-idealist conception of the human condition that seeks to displace the traditional definition of *animal rationale* with the notion of *animal laborans* (Arendt 1998: 105). At the same time, however, labour as 'man's metabolism with nature' designates for Marx a realm of necessity to be overcome in favour of a 'realm of freedom' where the condition of labouring for survival is surpassed universally (Arendt 1998: 104). Arendt has two main objections to this vision of social-political redemption: first, it suggests that labour required for human biological survival can be ultimately dispensed with, and, secondly, it removes the 'worldliness' granted to human existence through work. As she puts the latter point: 'The ideals of *homo faber*, the fabricator of the world, which are permanence, stability, and durability, have been sacrificed to abundance, the ideal of the *animal laborans*' (Arendt 1998: 126).

To this argument Arendt adds an intriguing element of social criticism: Marx and his fellow travellers in the nineteenth-century workers' movement were guilty of naivety when they projected the creative potential that would be realised by the masses once the burdens of labour where taken away. Against this Arendt decries the state of popular culture in language that recalls Horkheimer and Adorno in their celebrated 1944 publication *Dialectic of Enlightenment* (Adorno and Horkheimer 1997 [1944]): 'A hundred years after Marx we know the fallacy of this reasoning; the spare time of the *animal laborans* is never spent on anything but consumption, and the more time left to him, the greedier and more craving his appetites' (Arendt 1998: 133). The emergence of full-blown consumer society brings with it two principal evils: first, a boundless sphere of desire that can never be satisfied and threatens to use up all available resources; secondly, the eclipse of the public sphere due to the erosion of all that endures in the common human 'world' in favour of the fleeting quotidian needs of human labour and consumption. As Arendt eloquently makes the latter point:

Only the existence of a public realm and the world's subsequent transformation into a community of things which gathers men together and relates them to each other depends entirely on permanence. If the world is to contain a public space, it cannot be erected for one generation and planned for the living only; it must transcend the life-span of mortal men. Without this transcendence into a potential earthly immortality, no politics, strictly speaking, no common world and no public realm, is possible. (Arendt 1998: 55)

It is striking how readily Arendt's appeal to intergenerational responsibility here can be directly related to the dominant concept of sustainability. That concept, in the canonical formulation of the United Nation's 1987 report *Our Common Future* (the Brundtland Report), defines sustainability as an effort to meet the present generation's needs without compromising the ability of future generations to meet their needs. The report's focus on meeting the basic needs of the poorest global populations through western-style economic development, by contrast, firmly places it within the labour dimension of human existence. The fact that sustainable development is essentially about distributing goods in ways that keep a growing population alive, means that the qualitative elements central to what Arendt regards as higher, political life have been left out of account. Had she been around to read it, Arendt's reaction to the Brundtland Report would no doubt have followed her criticism of utilitarian thinking on questions of value. Such a framework, she states, 'gets caught up in the unending chain of means and ends . . . in other words, utility established as meaning generates meaninglessness' (Arendt 1998: 154). Interestingly, Arendt sees an alignment here with the anthropocentrism of Kant's alternate ethical paradigm. For in Kant, she affirms, human ends justify the drive 'to degrade nature and the world into mere means, robbing both of their independent dignity' (Arendt 1998: 156).

Contrary to initial impressions, Arendt is not here condemning work as the root of all ecological evil. It is rather the subordination of work to labour, that is, work performed solely for the sake of human consumption, that she criticises. This means that work, rather than erecting the permanent setting of common life in the form of truly public space, is directed towards the transitory satisfaction of mass society. It is therefore a reconfiguration or narrowing of work that concerns Arendt, a readily apparent phenomenon that furnishes one basis for the manifest unsustainability of modern global production. For Arendt, Marx was also wrong to see the mechanisms of the global marketplace as intrinsically destructive of the social value of labour and work. On the contrary, Arendt insists, it is precisely this marketplace in terms of the 'public realm' that bestows value on anything in the first place. As she puts it: 'Value is the quality a thing can never possess in privacy but acquires automatically the moment it appears in public' (Arendt 1998: 164). Marx's mistake, Arendt continues, was his insistence on seeing human labour as a reliable measure of value. In reality, there is no such measure whereby value can be fixed, as it is always subject to the vagaries of social valuation. Putting society on a firmer footing, for Arendt, does not come from reducing work to labour but rather from referring both to political action (in Marxist terms, praxis) within its proper theatre of the public sphere.

NEOLIBERALISM AND SURVIVAL POLITICS

According to Marx, when it comes to its management of the workforce there is only one limit recognised by the capitalist exploitation of living labour: mere existence. This or that individual worker can, of course, be quite literally worked to death. But some other living body must be found as replacement. The realities of this logic are currently being forced on the workers of affluent countries through the ongoing austerity politics. It is a measure of neoliberalism's success that political parties across contemporary liberal democracies do not pursue the social question of the value of work to the average worker. When the transition to the consumer society was in full swing in the 1950s, by contrast, this question was at the centre of debate. The examples given of influential political and social theory written in this period serve to demonstrate the focal significance of questions relating to the nature of work at the dawn of full-blown western consumerism. The important point for the current analysis is how these earlier social critiques commonly see the key problem as the proper relationship between work and public space. Whether in Arendt's analysis of the ancient Greek polis as the sphere in which shared life beyond biological life is forged, in Marcuse's indictment of modern work as inherently repressive, or in Lefebvre's conception of the right to the city, there is a clear conjunction of the problem of work with the quality of urban life.

But how does all this more explicitly relate back to contemporary debates on sustainable development? When we look at the contemporary problem of climate change in light of this earlier debate we notice something very striking: the neoliberal concept of sustainability remains oblivious to the very problem of alienated work. Embodying as it does an essentially neo-Malthusian appreciation of the human/nature relation, contemporary sustainability is driven by concern for the basic biological needs of the human population. In other words, sustainability is at base a survival politics. As such it remains aloof from questions about the quality of life, human or otherwise. This allows the sustainability paradigm to sidestep those issues of work that relate to meaningfulness and qualitative human flourishing. Once the environment is grasped as a means to ensure mere human survival, the only question that remains is the technocratic one of how to manage the productive apparatus to keep alive the 10 or 11 billion people the UN predicts will be living on the planet by mid-century.

Measured against the index of biological survival, past and current debates concerning the social value of capitalist production are of no relevance to environmental policy. For the same reason, when environmental catastrophism is embraced by political progressives they adopt the very terms of debate established by neoliberal governance. In the minds of those who consider environmental challenges to be fundamentally scientific and their solutions essentially technical in kind, this apparent lack of a political dimension is no doubt welcome. But this is mere appearance. As I argue throughout this book, the sustainability paradigm is not politically neutral. Rather, it is precisely designed to facilitate a neoliberal approach to the entire field of environmentalism.

As discussed in Chapter 1, neoliberalism exhibits an underlying naturalisation of human society thanks to the market being grasped as an eminent expression of human nature. Under neoliberalism, not only is the market the sphere that allows human nature to be expressed, environmentalism is also reduced to the scientific management of natural resources. Incorporated into the economic growth paradigm, the guiding question of sustainability becomes how quickly we can devise technologies to maximise resource extraction, distribution and usage. Thanks to our analysis of such thinkers as Smith and Latour, by contrast, it was possible to frame the central question of political ecology quite differently: How can human flourishing be pursued within the context of a general flourishing of all life forms? While the neoliberal paradigm reduces political action to human resource management and ecology to natural resource exploitation, the more critical perspective advocated here identifies crucial intersections between the question of non-alienated labour and non-exploited nature.

For all its neo-Malthusian talk of scarcity, the neoliberal ruling passion for unlimited economic growth in effect denies that human flourishing relates to a limited material foundation. Contrary to the flat-world neoliberal vision, this material foundation is never uniform but instead always involves specific characteristics of this or that concrete place. Consequently, all environmental politics is confronted with concrete problems of place-based difference. Neoliberalism has a starkly inconsistent appreciation of this. On the one hand, it proliferates images of urban uniqueness in parallel with the logic that advertises mass-produced commodities as personalised invitations to consumption. On the other hand, it produces a real world of smooth boundaries, where the affluent consumer can freely transit a seemingly unending environment of consumption. In face of this, it is important to remain cognisant of the realities of starkly uneven development that capitalism is intrinsically incapable of moving beyond.

Beyond the theoretical handwringing about how difficult it is to determine the third pillar of social sustainability relative to the other pillars of economic and environmental sustainability, clearly the sustainability paradigm in general is little more than a bureaucratic, ideological elaboration of neoliberal economics. The basic point here is that the sustainability agenda works by suppressing questions relating to the social pathologies that characterise neoliberalised labour and nature. Thus, the entire legacy of social critique expressed through the keyword 'alienation' is held in abeyance. Social policy penalising those who are not in paid employment was an early and familiar strategy deployed by neoliberal governance in the 1980s and 1990s. Here, the Situationist admonition articulated by the slogan 'never work' has become a mark of shame attached to the socially dispensable. But how can a serious discussion of social sustainability be conducted without the nature of contemporary work being up for debate?

My further point here is that environmental issues more generally cannot be tackled without the question of work (the problem of 'human resources') being central to such discussions. All these dimensions of social theory were crucial to wider debate in the 1950s and 1960s and yet now they constitute at best a subgenre of academic sociology. So successful has the neoliberal social construction of employment scarcity become. Ultimately, tackling environmental sustainability requires that we return to the question of socially necessary production, on the terms expressed by Smith (2008). Rather than givens, as neoliberal economics insists, human needs are in large part cultural and social constructs. There is a profound parallel between the neoliberal construction of employment scarcity and the neoliberal presentation of sustainability in terms of the maximised efficiency of material production. The latter is used as justification for the former. Neoliberalism's production mantra echoes the famous first words of Andrew Marvell's poem 'To his coy mistress': 'Had we but world enough, and time . . .'. But in the mouth of the neoliberal proponent these words do not offer a meditation on mortality but rather a specific political construction of environmental scarcity. The first step in overcoming neoliberal environmentalism is to strip away the veneer of neoliberal reality. As Harvey (1996) remarks more generally in connection with environmental thinking: 'Ideas about environment, population, and resources are not neutral. They are political in origin and have political effects' (Harvey 1996: 148).

The central question addressed in this book is how climate change relates to these patterns of neoliberal sustainable development. Not only can climate change not be tackled on the terms of such development, the neoliberal SD configuration should be seen as a crucially salient symptom of our present environmental situation. As the effects of climate change sharpen, so the social oppression issuing from neoliberal governance will become the more draconian and evident. The mantra of this political regime is: security, security, security. Climate change might be better labelled climate insecurity. The interactions within the global climate are so bewilderingly complex that the climate experts themselves readily admit that accurately predicting the pace and severity of change is beyond them. What we can say with certainty, however, is that climate change will be used to tighten the social screws of the neoliberal security agenda. As Welzer (2012) chillingly remarks: 'All the historical evidence makes it highly probable that "superfluous" people who seem to threaten those enjoying relative prosperity and security will lose their lives in increasingly large numbers, whether from lack of food and clean water, from frontier wars and interstate conflicts resulting from changed environmental conditions' (Welzer 2012: 181).

But the same logic of human superfluity can be observed within the world's wealthiest nations. Rather than a natural disaster, Welzer (2012) notes, Hurricane Katrina represented 'the creation of a new category of refugee and a new social demography of the city: the whole concatenation of events would be much more accurately described as a *social disaster*' (Welzer 2012: 25). As the stark events of post-Katrina New Orleans graphically demonstrated, neoliberalism always stands ready to take advantage of an environmental crisis to transform a city more quickly in accordance with its image. In the case of New Orleans the devastation was used to fundamentally re-engineer the public school system at a pace that would otherwise never have been possible. It is thus clear that the prospect of ecological catastrophe is something that will aid rather than undermine neoliberalism.

Neoliberal governance will become more draconian and socially repressive as climate change worsens. It will justify these developments in the name of human survival (with the obligatory rhetoric of safeguarding the 'free world', and so forth) and become increasingly intolerant of all collective action that stands opposed to its vision of social development. It will assure us that all possible resources are being harnessed to find solutions to climate change. While indulging in populist rhetoric it will insist on the message that government and business are hard at work to protect people from the harmful effects of climate change. In other words, climate change in the age of neoliberal governance will amount to a permanent state of emergency across wealthy liberal democracies, the professed bastions of collective liberty and individual human rights.

NEOLIBERALISM AND THE URBAN ENVIRONMENT

The central argument offered in this book is that climate change is a symptom of neoliberal governance that cannot be overcome, or even accurately diagnosed, from within this political regime. The concept of sustainable development is a quintessentially neoliberal notion. Consequently, the sustainability agenda must be abandoned in favour of quite a different approach to political environmental thinking and action. The main point added to the critique in this chapter is that SD views the purpose of work and production as biological human survival and thereby fails to engage with concerns about the social and ecology quality of work. These concerns are crucial because the essence of environmental health is the flourishing of the entire system of life. Obscuring the question of the social value of work in favour of inexorable economic growth in the name of human survival is an intrinsic and ineradicable trait of neoliberal governance.

Admittedly, the neoliberal paradigm has also given rise to a happiness industry recently analysed as the 'wellness syndrome' (see Cederström and Spicer 2015). But the appeal to individual self-cultivation is done in the name of social competition rather than cooperative solidarity. As recent studies have shown, many large employers use wellness programmes to discipline their employers in the name of company productivity. Such innovations are very far from the kinds of free cooperative action envisaged by the earlier social theory explored in this chapter. In their different ways, thinkers such as Lefebvre, Marcuse, and

Arendt could agree, almost six decades ago, that the most pressing questions related to the social value of work. Neoliberalisation has eclipsed this concern in favour of a one-dimensional growth agenda in global politics. Social development, considered without qualitative markers, is uncritically regarded as a given by-product of economic growth.

With a severely restricted field of political good sense, we find ourselves in the preposterous position of looking to neoliberalism to cure us of the very condition it has engendered. But the antidote in this case cannot be derived from the poison. Part of the neoliberal conceit here is precisely to deflect attention away from its pivotal role in the cause of climate change in favour of its ability to produce solutions. This also means that the social and political significance of climate change is sidelined and the problem cast as a technical natural resource issue. As Smith (2008) remarks:

> Insofar as global warming as a process is taken in isolation as the central environmental dilemma – and is thereby extracted from the processes of capital accumulation and the social relations of production which significantly provoke such climate change – the dynamics leading to global warming fall out of focus. (Smith 2008: 247)

The pathologies of neoliberal economic development are nowhere more graphically displayed than in contemporary cities. Certain patterns of urbanisation are a necessary feature of neoliberal development, whether this development is considered 'sustainable' or not. The slum cities of the developing world and the gentrified urban landscapes of developed nations have this in common: they are ultimately symptoms of the fundamental drive to intensify rent extraction. Depopulation of the countryside in developing nations conveniently removes human occupants whose lifestyles represent blockages to the neoliberal goal of maximised profit extraction. In developed economies, the mounting pressure on the lives of workers faced with runaway growth in real estate markets and increased insecurity of employment chances and conditions has given rise to periodic urban protest and occupation movements.

Despite populist rhetoric, governments across liberal democracies go to great lengths to have their major cities increasingly dominated by international financial capital and property developers. For as long as the spectre of an alternative political reality existed within the Soviet bloc, liberal democracies had something to prove. After the collapse of the Soviet Union and the ecstatic proclamations of the end of history, neoliberalism could intensify its reshaping of society safe in the thought that its chief political challenger had now been permanently discredited. The age of totalitarian oppression was over and the 'open society' could now expand unopposed. But neoliberal freedom in reality took the form of a world increasingly shaped by the interests of transnational capital. On the national level, liberal democratic politics increasingly focused on free trade agreements; at the municipal scale the cities of these nations were socially cleansed of all activities other than those focused on consumption. While domestic space was colonised by inexorable waves of real estate speculation, ostensibly public space was being converted into privately owned and managed retail landscapes.

Traditionally suspected of breeding vice and corruption, the city under neoliberal control was being converted into a place of innocent, even virtuous consumption. The greening of western cities is a complementary palliative of this urbanisation. Thanks to advanced globalisation, western cities were no longer obliged to house industrial activity, and so the remnants of the industrialised landscape were drawn into the neoliberal property market. This 'urban renaissance' is part of a much more generalised neoliberal process of sanitising the industrial past in order to make it ripe for further waves of popular consumption (see Hatherley 2016). Here, the general public is quite literally invited to gorge themselves on a carefully repackaged version of their own social history. In the UK, for example, the docks of Liverpool and the canals of Manchester became icons of post-industrial development. Initially slower than its textile-manufacturing neighbour, Liverpool's riverfront has recently been materially reimagined in the shape of the gigantic retail funhouse known as 'Liverpool 1'. Following decades of decline, Liverpool's population has started to grow again. Given this belated but glorious urban revival, no one is expected to raise critical questions about the quality of the jobs created or the reality of the personal debt needed to popularise the growth in consumption. It is enough to see that the people have enhanced opportunities to shop.

Recent protracted social upheavals in cities across the world have made clear that there are cracks in the highly polished veneer of this neoliberal consumer utopia. The Occupy movement underscored the fact that the urban environment is the most likely site of resistance to neoliberalism. When people take to the streets, apparently demonstrating nothing but the public's right to occupy public space, neoliberal governance is forced to walk a tightrope. If security measures, particularly in liberal democracies, are seen to be too obviously brutal and violent, this tears holes in the carefully managed stage-set of social beneficence that neoliberal capitalism, despite everything, is required to display. On the other hand, a central task during the era of neoliberalism has been the political and social sanitising of the urban environment. 'Zero tolerance' policing has socially cleansed western cities, allowing the real estate and retail sectors to triumphantly recreate the city in their own image.

Meanwhile, the realities of precarious urban work have made living in the central city less and less viable for large swathes of the urban population. Public authorities and institutions, notably urban colleges and universities, have been eager agents in this process. While planners congratulate themselves for repopulating city centres that had become deserted business districts outside regular office hours, urban residential divides across liberal democracies have once again come to take on the stark contours observed by Friedrich Engels in mid-nineteenth-century England. Neoliberal economics necessarily works through the dominance of the financial sector relative to manufacturing. This shift in balance has been crucial in undermining the power of organised labour. The fact that consumer goods are overwhelmingly produced beyond the wealthy consumer nations also ensures that awareness of the environmental and social harms of contemporary production are little more than an occasional media image to consumers.

What cannot be so readily hidden from view is the social harm done by neoliberal urban development. The requirement to construct 'affordable' housing is routinely flouted by developers with the connivance of city officials. In the United States, the use of tax increment financing (TIF) creates urban renewal areas that effectively divert huge amounts of property tax revenue away from essential public services such as schools. Other public institutions such as libraries are often kept afloat through private donations or additional local taxes agreed to by urban residents. Another sign of the dominance of big money, city officials increasingly agree to large levels of public subvention for the construction of private sports facilities. In these different ways neoliberal governance is presiding over the rapid erosion or disappearance of public space.

In contrast to the panic sown and perpetuated in the face of climate change, the task of undoing neoliberalism must be measured and painstaking. Just as the features of neoliberal development are most strikingly on view in the large city, so too is the metropolis the most likely site of collective acts of resistance against it. Neoliberalism has by now grown such deep roots as a common sense social imaginary that the first crucial task is to openly contest this constriction of collective imagination. It must be demonstrated that neoliberal economic development is not the inevitable and only means of human progress. In this light, the achievement of the Occupy movement was its very occupancy of public space for the sake of open political debate. Faced with this, the neoliberal urban security agenda was obliged to become more visibly oppressive than it prefers to be. The spectacle of violently removing such offensive structures as a spontaneous people's library made clear the absolute intolerance neoliberal urbanism harbours towards all elements of non-consumption within the city.

The point made was dramatically anti-democratic and amounted to saying: you, the people, have no eligibility as a political actor. It embodied what the contemporary French thinker Jacques Rancière (2014 [2005]) calls straightforwardly 'hatred of democracy'. As the Situationists clearly articulated in the 1950s, the modern western city is capitalist ideology made material in the form of the urban environment. Once an ideology becomes material it is more than a mere mental framework. To return to the game metaphor used in Chapter 1, the neoliberal city offers a horizon of action in which only certain moves will be deemed socially legitimate. According to the traditional liberal regime of rights, the Occupy encampments should count as a manifestation of the freedom of assembly. While the rapidity and spontaneity of the Occupy movement temporarily put neoliberal urban governance on the back foot, their protracted nature allowed the encampments to be condemned as illegitimate trespassing on public space. The paternalistic message was clear: all right, children, you've made your point; now it's time to let the adults take over again.

The value of Occupy was that it drew attention to the fact that much of what is presented as public space in the neoliberal city is only publicly accessible on very restrictive terms. Spontaneous occupation by the public is the last thing the private companies who manage much contemporary 'public' space are likely to welcome. This is for the simple reason that there is no profit to be made by such occupancy. Just as the neoliberal environmental agenda is dominated by the issue of natural resource security, so too is neoliberal governance preoccupied by an effort to keep all non-consumption-based activities out of the urban environment.

The right to the city notion advances the claim that democratic society has halted at a crucial stage in its evolution. The process of realising a condition of universal democratic enfranchisement in the nineteenth and early twentieth centuries was meant to render political decision-making truly reflective of popular concerns. Universal, state-funded primary and secondary education was meant to facilitate a well-informed and judicious electorate. But once these preconditions were met, along with improvements to worker conditions through unionisation, the march of democratic popular self-determination essentially came to a halt. Economic prosperity was more and more offered as an alternative to real political and social empowerment.

As Adorno and Horkheimer (1997 [1944]) could already discern in 1940s America, the 'culture industry' was rapidly anaesthetising popular political consciousness, relaxing the social tensions that had led workers to call for change through the soporifics of consumption and entertainment. While Walter Benjamin (2007 [1955]: 234) could discern in Chaplin's 1936 film *Modern Times* an effort to focus popular discontent within the mechanised workplace, by the 1940s and 1950s the Hollywood machine had been successfully converted to full-scale production on behalf of a resurgent global capitalism. In our own times, the constant panegyrics for new social media seem oblivious to Debord's (1994 [1967]) warnings about the 'society of the spectacle' issued almost half a century ago. Contrary to the common contention, the political significance of the Occupy movement (inspired by the 'Arab spring') does not reside in the novel deployment of social media but rather in the very occupying of ostensibly public space by the public. In a putatively post-ideological world where the spectres of political alternatives to neoliberal capitalism have been all but exorcised from popular consciousness, 'the people' have no political tasks left to them other than that of obedient over-consumption. If a genuinely democratic politics is to be reinvented this will unavoidably involve reimagining the urban condition. In this sense, at least, the most intensely neoliberalised landscapes can be regarded at the most politically promising.

In conclusion, actually existing urban sustainability should be seen as an ideological outgrowth of mature neoliberal economic hegemony. Its success is an important part of the gradual liquidation of the more radical elements that characterised the environmental movement in the 1960s and 1970s. Belated efforts to provide a 'spatial fix' for lack of equity in urban environments are mere epiphenomena that, at best, offer a facade of political progressiveness to what are otherwise socially regressive configurations of the city. On the terms it has operated by thus far, it is high time for the urban sustainability experiment to be declared the social failure it undoubtedly is. But this will only be a productive move if we return to the debates of the 1950s that began to think through the negative social and political impacts of mass consumption. On this score, the recent re-emergence of the notion of a right to the city offers a useful framework for a very different configuration of genuine collective urban flourishing.

The confrontation with neoliberal urbanism poses a fundamental question: what is the social purpose of the city? Historically, the first cities were made possible by the establishment of permanent agriculture. In this sense it can be said that in their very origins cities are a crucial manifestation of what Marx called the metabolic relationship between 'man and nature' (see Foster 2013). Urban sustainability as it is currently conceived and practised has all but forgotten the crucial dialectic between the city and the country. The western neoliberal city has achieved a remarkable dematerialisation over the last four decades, as manufacturing has given way to the service and financial sectors. It has also, equally remarkably, led to the virtual absence of the urban crowd dedicated to anything other than consumption. In this context the ongoing 'revolution' in communication technology does not reverse but rather exacerbates the socially negative impacts involved in the disappearance of the urban crowd. Once again, the Situationists were prophetic on this point: the capitalist city has increasingly drawn the urban population back to the confines of its domestic environment, thereby rendering the reality of public space merely notional. The point here is not how many bodies one can count on the urban street, but the social meaning those bodies constitute. One might say, with only a little hyperbole, that there are in the neoliberal city only two forms of urban presence: legitimate consumption or illegitimate non-consumption. The right to the city asserts first and foremost that the presence of non-consumers on the street in not merely legitimate. It is the very act that defines the socially vital urban environment.

CHAPTER 5

The city and the country: towards a new environmentalism

> So if the urban process is open-ended and if urbanisation is global and boundless, any transformative politics presumably need to be likewise. (Merrifield 2013: 916)

THE CITY AND THE COUNTRY

The previous chapter brought together key elements in our analysis of climate change through a consideration of work and urbanisation under neoliberalism. 'Development' more generally, and therefore sustainable development more particularly, intrinsically involves urbanisation. But urbanisation does not have any natural or inevitable form, as attested to by the diversity of visions for the good urban life projected in ancient and modern times. As asserted by Lefebvre (2003 [1970]) in the wake of the 1968 events, the capitalist urban condition is not synonymous with the city. As noted in our discussion of Arendt, using the ancient Greek polis as a model for urban existence under advanced capitalism is illegitimate. The polis was a geographically limited site with a typical population of no more than 100,000. Of this population perhaps one in five would constitute politically enfranchised individuals in a democratic polity (see Gates 2011). Political life would therefore consist of regular encounters with a large but strictly encompassed class involved in the *bios politikos* (political life).

Thus, while the polis arguably constituted a genuine political *Gemeinschaft* (community), the modern city corresponds to a *Gesellschaft* (society). In the latter case face-to-face political community is untenable, thereby necessitating various communication media to establish the 'imagined community' (Anderson 2006 [1983]) of the modern nation state. The industrial metropolis and, *a fortiori*, the neoliberal megalopolis, is not a readily delimited site. Rather, virtual boundlessness is its distinctive feature. This very sense of limitlessness no doubt accounts in part for the stratifications and striations of modern urban society. Where there are no outer bounds, inner boundaries spring up more frequently and emphatically. Nineteenth-century accounts of a city such as London abound in metaphors indicating a basic inability to encompass the modern city (see Ackroyd 2000: 562–85).

When insisting that the urban condition has become general, Lefebvre was not simply repeating what was by then a well-established cliché about the massiveness of the modern city. His underlying point is that the quality of the urban has come to qualify all real and potential places under capitalist globalisation. The most obvious evidence for this is the fact that mass media operations, while they tend to be physically situated in cities, generate content that is disseminated throughout all, or most, rural locations. As noted, environmentalism's preoccupation with wilderness can readily be shown to grow out of a romantic sensibility characterised by moral condemnation of urban existence. While such existence is portrayed in terms of sophistication, superficiality and luxury, to rural life the corresponding virtues of simplicity, profundity and thrift are attributed. On the foundation of these oppositions a critique of bourgeois existence is constructed running from writers such as Rousseau, Herder and Goethe to Blake, Wordsworth and later Ruskin. In his seminal study *The Country and the City* Raymond Williams (1973) captures this anti-urban bias in the following way:

> Blake saw a common condition of 'weakness and woe'. Wordsworth saw strangeness. A loss of connection, not at first in social but in perceptual ways: a failure of identity in the crowd of others which worked back to loss of identity in the self, and then, in these ways, a loss of society itself, its overcoming and

replacement by a procession of images: the 'dance of colours, lights, and forms', 'face after face' and there are no other laws. No experience has been more central in the subsequent litera- ture of the city. (Williams 1973: 150)

But Williams (1973) immediately adds to this experience of urban social dissolution a counterpoise of solidarity and coordi- nation appearing as a kind of obverse, redemptive image:

> This historically liberating insight, of new kinds of possible order, new kinds of human unity, in the transforming experience of a city, appeared, significantly, in the shock of recognition of a new dimension which had produced the more familiar subjective recoil. The objective uniting and liberating forces were seen in the same activity as the forces of threat, confusion and loss of identity. And this was how, through the next century and a half, the increasingly dominant fact of the city was to be paradoxically and alternatively interpreted. (Williams 1973: 151)

Williams' basic idea here is that poetic evocations of urban experience within the early industrialising city by the likes of Blake and Wordsworth point to a profound internal tension. On the one hand, the modern metropolis is seen as a loss of some supposed rural simplicity and sense of community; on the other hand, the very massing of individuals within the com- plex fabric of mechanical production allows one to sense the potential for a more profound form of human connectedness and cooperation. This latter intuition is, of course, basic to the socialist critique of capitalist exploitation. Involving more intricate networks within the division of labour, industriali- sation in fact makes individuals far more dependent on each other relative to simpler agrarian societies. Yet that dependence is predominantly forced into the mould of mutual competition. This framework in turn arises from the basic fact of capitalist restriction of ownership. Where a minority claims the wealth of industrialisation, the majority are turned over to an unrelenting struggle for existence.

Following Williams's (1973) analysis, it is important to note that any cogent critique of neoliberal environmentalism should steer clear of the anti-urban bias inherited by much mainstream environmental thinking. This is no way implies that the complacent neoliberal celebrations of contemporary urban 'revitalisation' should pass unchallenged, but rather that the shortcomings of the urban cannot be credibly overcome through appeals to the rural or wild as a counter-urban antidote. After all, Lefebvre's basic point is that the urban becomes ubiquitous – equally qualifying rural locales – under advanced capitalism. It is obviously not an accident of history that, with the growth of the giant industrial cities in the nineteenth and twentieth centuries, agricultural practices became intensively mechanised in the search for higher yields. To repeat: it is not a matter of joining the chorus denouncing the evils of urban life, but rather a question of undermining the seeming inevitability of the urban in its neoliberal modulation.

This sounds like a rather vague task, but in fact it takes quite concrete forms. To name the most obvious, the very occupation of nominally public space by the public contests the neoliberal consumption imperative. Currently, the only unquestionably legitimate instances where large assemblies of individuals are permitted in liberal democracies relate to sport and music events. Given the relative ease of access to such spectacles offered by contemporary communications media it is surprising that so many are willing to spend large sums to be physically present at the live event. There is no doubt an element of class appropriation going on here, with formerly working-class sports being increasingly followed by the middle class (see Jones 2011: 134–42). Equally, the youth culture connected to music can be more lucratively exploited when the not so young with greater financial means start attending music concerts and festivals.

Below the surface, however, we might discern a simpler, more general motive at work: the desire to be part of a crowd. It is easy to forget how much nineteenth-century political discussion was dominated by fear of the public mass or working-class crowd. The idea that the 'lower orders' were incapable of appearing in the public domain without a tendency to social disturbance and property destruction was a common prejudice. Matthew Arnold's account of the 1866 Hyde Park events in his *Culture and Anarchy* is a classic of the genre (see Williams 2005 [1980]). In our supposedly more egalitarian times it is telling how readily this presumption of the unruly public mass bursts forth once again. No doubt studies of the mass psychology of crowds that sought to shed light on the sociology of totalitarianism (see Canetti 1984 [1960]) added a veneer of scientific respectability to the traditional prejudice against the massed public. The prejudice was certainly on proud display in much mainstream media coverage of the Occupy protests. It is as though Plato's image of the demos as an untamed wild beast only to be kept in check by the discipline of a wise master still inhabits our supposedly democratised collective imagination.

In all social-historical analysis, as again Williams points out, it is important to disentangle archetypal tendencies of thought from the specific material conditions in which they are applied. For us this means that it is the specific animus against the appearance of the public in the neoliberal urban context that needs to be understood. Initially, we can ask: what does it say about the condition of contemporary liberal democracies that the mere appearance of large numbers of people in public space amounts to a political provocation?

THE URBAN EXPERIENCE AND MILITANT PARTICULARISM

Harvey (1996) takes up the concept of militant particularism from the work of Williams and defines it as a process in which '[i]deals forged out of the affirmative experience of solidarities in one place get generalised and universalised as a working model of a new form of society that will benefit all of humanity' (Harvey 1996: 83). Williams's personal point of reference is his experience growing up in a Welsh border town dominated by the coal industry. Harvey's own point of departure is his involvement in a campaign to save the Cowley car manufacturing plant in Oxford. The problem, analytically, relates to the universalist vision of Marxist revolutionary socialism. Harvey (1996) quotes the following key passage from Williams:

> A new theory of socialism must now centrally involve *place*. Remember the argument was that the proletariat had no country, the factor which differentiated it from the property-owning classes. But *place* has been shown to be a crucial element in the bonding process – more so perhaps for the working class than the capital-owning classes – by the explosion of the international economy and the destructive effects of deindustrialisation upon old communities. When capital has moved on, the importance of place is more clearly revealed. (Harvey 1996: 80)

Williams's sense that place would become a central political preoccupation in the decades to come was acutely prescient. Of course, the way in which that turn back to place has been shaped by the forces of neoliberal capitalism is quite opposed to the kind of solidarity Williams had in mind. Neoliberal urbanisation was arguably given its first canonical formulation by the New Urbanism movement founded in the 1980s (see Haas 2008). At first going by the name of 'neotraditionalism', New Urbanism sought to design community back into life by means of traditional building types and public space. In reality, however, New Urbanist developments offered places of retreat to the upwardly mobile, thereby offering yet another historical variation on the theme of the rural or suburban villa. As New Urbanism has largely focused on greenfield sites, it was never going to offer a credible neoliberal model for truly urban 'regeneration'.

This task has instead by assumed by metropolitan governments, which have created urban renewal areas (URAs) that divert large amounts of local taxation from public infrastructure such as schools. The diverted funds in effect pool resources so that developers can be relieved of the burden of making improvements around what they build. Given that URAs typically last for a generation they work to intensify uneven urban development and help to undermine truly public institutions by starving them of funds. While the typical rationale for establishing an URA is some initial claim of underdevelopment, metropolitan governments also like to use them to showcase green infrastructure such as low-energy buildings, storm-water mitigation and enhanced pedestrian and cycling infrastructure.

It is clear, however, that such measures do not represent a genuine effort to tackle environmental quality within the city as a whole. Instead, URAs create green islands that simply add an ecological dimension to other, more readily apparent aspects of uneven development. In this way, neoliberal environmental urbanism makes an important contribution to the green consumerism paradigm. According to this paradigm the best way for the average citizen to display environmental agency is through acts of consumption. Buying in to a green neighbourhood can thereby be presented as both a canny economic investment and a sound environmental ethical decision. The logic of green consumerism can also be conceived of as a general project of environmental pacification, that is, as an effort to mask the social contradictions and tensions that inevitably arise in the city under neoliberal governance. For all its claims to enhance urban life, neoliberal environmentalism can no sooner produce a genuine urban commons than neoliberal finance can allow for equal access to credit. Trickle-down environmentalism is no more convincing than trickle-down economics.

Despite all the shortcomings of neoliberal urban environmentalism, it would be foolish to deny that it has been remarkably successful in appealing to a certain kind of attachment to place. Environmental localism, as frequently noted, has consistently represented a key affective dimension of the environmental movement more generally. Attachment to place undoubtedly plays a key role in individuals' sense of themselves: places remembered, lived, and anticipated offer key reference points for constructing personal identity. Williams's notion of militant particularism is, however, something more specific than simply an individual's subjective sense of belonging. It is a question of working-class, intergenerational solidarity. As Williams (1989) put it in a late interview with Terry Eagleton: 'It's the infinite resilience, even deviousness, with which people have managed to persist in profoundly unfavourable conditions . . . A sense of a value has won its way through different kinds of oppression in different forms' (Williams 1989: 322).

Municipally orchestrated urban improvement is highly unlikely to constitute this kind of sense of community resilience for the simple reason that, to use Williams's language, this is the urbanism of the oppressors. Indeed, embattled working-class communities rightly see urban environmental improvement as simply another front in an inexorable process of gentrification. The green infrastructure is attractive to developers not because they harbour deep ecological commitments, but rather because it facilitates higher levels of rent extraction. This necessitates the displacement of the original population, which could reside in this or that neighbourhood precisely because rents and house prices were lower. In other words, as critical geographers have been pointing out for decades, capitalist urban regeneration is simply one aspect of a comprehensive process of uneven development.

The task, then, becomes one of clarifying how Williams's militant particularism can be used to conceptualise cogent critique of the neoliberal urban condition. If, as Williams contends, place remains or even grows in importance when capital withdraws, how can this collective attachment to place direct contemporary environmental practice and constitute credible opposition to neoliberal environmentalism? We have already touched on one possible answer to this question when considering Neil Smith's (1984) idea that the central problem relates to socially necessary production (Smith 2008). A tradition of critical thought, stretching from the work of the Frankfurt School in the 1940s to contemporary critical geography, has built up a consistent critique of consumption as social-political pacification. In the previous chapter we saw how, in the late 1950s, thinkers such as Marcuse and Arendt drew attention to the corrosive effect of consumerism on collective agency and the political domain more generally.

In its simplest form, the idea is that endemic commodity consumption, facilitated crucially by modern advertising and entertainment media, distract the public from its manipulation by the forces of capital and quell the desire to bring about collective change. The gains of earlier generations of workers to secure shorter working hours and safer workplace conditions are taken for granted and the worker retreats further and further into the domain of private consumption. Despite claims that new social media have reinvigorated political activism, the state of party politics across liberal democracies becomes more moribund with every electoral cycle. Eight years after the beginning of the Great Recession, little has been achieved in the wealthiest economies to restrict the political action of effectively unregulated finance. The political efficacy of the Occupy movement ceased almost at the moment the encampments were demolished by local police.

Considered broadly, the political situation is truly paradoxical. While environmentalists proclaim an imminent 'game over' situation for the planet, neoliberal hegemony shows no real signs of ceding to something more enlightened or genuinely democratic. Faced with destructive ecological change on a vast scale, systematic political change across wealthy liberal democracies has never seemed less likely. How is this possible? Fundamentally, I believe this is possible in large part due to the lack of a credible conception of environmental-political agency. The physical environment is generally taken to be something simply there, a kind of stage onto which we step to assume the role of actor. In other words, it is viewed as a sort of sterile container within which certain types of action can take place. This conception is a basic premise of the kind of Promethean humanism that sees nature as something to subjugate ruthlessly in order to satisfy human wants and desires.

Against this, as we have seen, Latour (2004 [1999]) argues for a more fundamental understanding according to which all agency is seen as involving human/non-human networks. This political ecology allows us to view social agency as basically environmental in kind. In this, Latour's perspective readily recalls one of the ecology's well-known mantras: everything is connected. As Harvey (1998) sardonically remarks, this may be true, but so too is the qualification: 'some things are more connected than others' (Harvey 1998: 329). Deep ecology's commitment to 'biocentric egalitarianism' (assuming all life forms to be of equal significance) undermines the very loyalty to place highlighted by Williams as a key resource for political struggle. For, if all forms of life possess equal 'intrinsic value', as the environmental philosophers like to say, then one's own sense of identity is dissolved into some supposed ecological whole. This is in fact the prescription of the originator of deep ecology, Arne Naess, who sees ecological agency as a function of an individual's ability to identify with ever-wider circles of the natural world (see Naess 1973). Here, losing an initially restricted sense of psychological selfhood is a step towards a genuine sense of ecological responsibility.

The basic point of militant particularism, by contrast, is that collective agency of resistance is predicated on a profound attachment to this or that concrete place. Thus, while the holism of deep ecology cannot be faulted on a purely theoretical level, its ability to provide the platform for critical environmental praxis is questionable. Deep ecology is modernist in the sense that its fundamental claim relates to self-realisation. It assumes, as an uncritical given, that the would-be environmental agent is not in a position to act. The notion of militant particularism assumes the opposite: that one is situated in a place whose historical character carries with it potential for resistant action. For deep ecology what blocks ecological agency is a faulty anthropocentric worldview that holds nature to be a merely passive tool of human manipulation. For Williams, by contrast, such agency is frustrated by capitalist development that inherently tends to use up all local resources, human and natural, before moving on to a new site of extraction. Williams sees environmental agency emerging from a radicalisation of partiality based on identification with a particular place; whereas Naess sees such particularity as a psychological failure to identify with nature as a whole.

For his part, Harvey (1996) attempts to make use of Williams's notion of militant particularism while remaining critical of what he sees as a questionable appeal to organic community determined by identity through attachment to place. His experience of the workers' efforts to keep the Cowley plant open convinced him that local struggles must remain mindful of the broader economic and political context. More generally, if socialist praxis were dissolved solely into various militant particularisms the sense of a wider collective struggle against a global capitalist consolidation of ownership would be lost. This was one basic reason for the internationalist orientation of the workers' movement as conceived by Marx. Accordingly, Harvey (1996) advocates a difficult balancing act:

> the politics of a supposedly unproblematic extension outwards from the plant of a prospective model of total social transformation is fundamentally flawed [. . .] Other levels and kinds of abstraction have to be deployed if socialism is to break out of its local bonds and become a viable alternative to capitalism as a working mode of production and social relations. But there is something equally problematic about imposing a politics guided by abstractions upon people who have given their lives and labour over many years in a particular way in a particular place. (Harvey 1996: 73)

Another way to capture the difference between this model of environmental agency and the green consumption model that guides neoliberal environmentalism is to say that the former draws upon the historical struggles of a resident population, while the latter appeals to a indeterminate class of ideally placeless individuals. Just like the International Style of architectural modernism, contemporary environmental urbanism offers homogenised place-less neighbourhoods presented as impersonal utopias of green consumption. Of course, advertising relentlessly resorts to artificial 'place-making' where strongly sanitised versions of local urban history are drawn upon or largely fabricated to give a sense of freewheeling, pseudo-personalised quirkiness. In reality, as many long-term local residents readily note, whatever genuine character is possessed by the pre-development site is quickly covered up so as not to alarm would-be investors that buying into the neighbourhood might be economically too risky.

The most obvious contradiction in all this is the fact that the natural resources required to transform a poorer city neighbourhood into a green showcase, combined with the intensified consumption within the post-development site, amount to anything but an environmentally positive transformation. But this is beside the point when it comes to the real logic of neoliberal environmentalism, which always comes down to consuming the image of environmental agency while destroying the material conditions for genuinely collective ecological action.

FROM THE TRAGIC TO THE COMEDIC COMMONS

In Harvey's recent work Lefebvre's idea of the right to the city has been brought together with the idea of a new urban commons (see Harvey 2012: 67–88). As a social-historical entity the commons represents land whose benefits are collectively, though restrictively available. In the context of feudal society it was a concession to commoners who were essentially disbarred from individual ownership. The features of the historical commons that are attractive to Harvey include its collective nature but also the fact that use-value rather than exchange-value lies at its core. Whereas the traditional commons centred on its agricultural use (typically a restricted right to graze animals or forage for useful material such as firewood), the modern urban commons would allow for spontaneous political action demonstrating self-organisation. Ostensibly, contemporary cities are already replete with public space for such purposes, but closer inspection makes clear that in many cases these spaces are privately owned and managed for the purposes of profit. Even where this is not the case there has been a well-documented tendency over recent decades for cities across liberal democracies to resort to heavy-handed policing in the case of spontaneous public occupation of public space (see Graham 2010).

Even though the right to assembly forms a classic element of the liberal democratic polity, legitimate practice of this right has been severely curtailed. This control of public space mirrors control of the workplace, which is by and large not something the typical worker has a meaningful part in. Once again we see here the familiar restriction of popular agency to acts of consumption. If the right to the city calls for popular determination of the urban fabric by the city's inhabitants, the creation of the urban commons allows for concrete places where the public appears to itself as a tangible political agency. The new urban commons would thus offer a site where a counter-neoliberal environmental agency could start to organise itself.

Of course, putting the matter in these terms makes clear that the urban commons will not simply be ceded by civic authorities but must be won through a protracted process of political struggle. This points to something of a paradox: in order to legitimate the appearance of the urban public within the commons this collective agency must first appear in a pointedly illegitimate fashion. This is what makes Harvey (2012) stress that the commons is, in social action terms, a constant process of collective 'commoning' (Harvey 2012: 73). All historic movements of resistance have understood that this process is inevitable. Defenders of the neoliberal orthodoxy will be bound to critique this call to create the urban commons. Public space, they will argue, already exists so why call for it to be created? If the public is allowed to mass on the streets, surely undesirable elements will take over? Hardworking members of the public will be too busy to show up, leaving the idle to present the face of the public. All these criticisms have been well rehearsed, from Matthew Arnold's Victorian England to contemporary media portrayals of the Occupy movement.

The construction of a commons brings with it the potential virtue of mutual trust and a public place of spontaneous, non-monetised celebration. If the more familiar concept of the tragedy of the commons refers to the supposedly inevitable exploitation of shared resources at the hands of self-interested individuals, what I am proposing here could be labelled the 'comedy of the commons'. The term is useful on a number of counts. First, it draws on the playful, spontaneous kind of activity celebrated in the Situationist attitude to urban experience. Second, like all good comedies, it allows frustrations and conflicts to occur, while tending towards a happy marriage in the final act. Thirdly, it allows for collective self-recognition of the common issues facing those who would otherwise retreat into a private scene of tragic isolation. The rituals of capitalist consumption, by contrast, take the form of individual repetitions that give only an abstract sense of social solidarity. Finally, like the ancient comedy, the commons offers a sort of social interlude, an interruption of the inexorable cycles of productive work. In this way, the commons represents a collective transcending of biological necessity that Arendt identified with the work of art. Only the work in question is not a material end-product but rather the event of celebratory gathering itself.

The commons as comedy is, then, a cyclical interruption in the weft of productive time in a particular, designated place within the urban fabric. It recalls, in all these ways, the traditional carnival, which similarly suspended the usual actions of production and pointedly disturbed the regularly social order. Unlike the soulless plaza dreamt up by urban planners within which computer-generated happy consumers are to be found, the new urban commons would be a genuine expression of popular agency. The environmental dimension would bear on the sense that this commons is as much a meeting place of non-humans as humans. Any truly democratic, post-neoliberal government, at whatever scale, could no sooner curtail or remove an urban commons than it could reintroduce a property qualification for enfranchisement. The commons is the people's place of appearance, where an unqualified mass demonstrates its potential for collective action. It foreshadows transition to the organisation of production by the associated workers, in other words, it points towards truly socialised satisfaction of human needs.

In an extensive consideration of the legal concept of 'inherently public property' from 1986, Carol Rose used the expression 'comedy of the commons' to refer to those customary uses of a space where open participation generates social value. Her point of reference is traditional festivals in England where use of a space was held open to a given community and legally protected from private rent extraction. Rose develops her analysis as an explicit counterpoint to Garret Hardin's concept of the tragedy of the commons. A key element in her account of the legal history is that customary social practice is the basis for the legal status of a commons:

> Custom thus suggests a means by which a 'commons' may be
> managed – a means different from exclusive ownership by either
> individuals or governments. The intriguing aspect of customary
> rights is that they vest property rights in groups that are indefinite

> and informal, yet nevertheless capable of self-management. Custom might be the medium through which such an informal group acts generally; thus, the community claiming customary rights was really not an 'unorganized' public at all. (Rose 1986: 742)

Hardin's 'tragedy' is predicated on the claim that any resource accessible by an indeterminate or 'unorganized' public would inevitably be subject to overuse and eventual exhaustion. Hardin's original application of the tragedy thesis was, of course, to environmental commons such as fish stocks in international waters. Key to Rose's alternative understanding is that a comedic commons is distinct from governmental legal claims such as those enacted under eminent domain, where a piece of land or real estate is bought by the state at market rates to realise a transportation plan or the like. In contrast, the customary commons is protected by the state but its use is decided by customary social use. That use is hedged with qualifications, just as the use of common land on the feudal manor was restricted by fluctuating conditions determining reasonable use.

Applying this idea to our case of a contemporary urban commons would amount to this: a recognised location legally protected by state or municipal power, the use of which is determined by customary usage. In some ways municipal parks are implicitly modelled on this notion of a commons. However, as recent Occupy events demonstrated, the neoliberal state readily overrides the claims of the commons in the name of a security agenda. But this simply makes clear the authoritarianism that can come to the surface whenever the public attempt to make a point of appearing in public space. The chief lesson of Occupy, seen in this light, is not the failure of the movement to create a common platform of political reform, but rather the insight that the customary practices on which a commons is based need to be essentially reinvented in the contemporary neoliberal city. As Rose (1986) puts it with respect to the comedy of the commons:

> Indeed, the real danger is that individuals may 'underinvest' in such activities, particularly at the outset. No one, after all, wants to be the first on the dancefloor, and in general, individuals engaging in such activities cannot capture for themselves the

full value that their participation brings to the entire group. Here indefinite numbers and expandability take on a special flavour, relating not to negotiation costs, but to what I call 'interactive' activities, where increasing participation *enhances* the value of the activity rather than diminishing it. (Rose 1986: 768)

Thus, the proverb capturing the spirit of the commons is 'the more the merrier'. This contrasts precisely with Hardin's tragic version, whereby the more seeking to take part the more misery is likely to ensue for all.

THE URBAN COMMONS AND ENVIRONMENTAL ACTION

The idea of the urban commons may seem quite removed from the environmental challenges involved in climate change. Predominantly, as we have seen, climate change is presented by means of a well-rehearsed script emphasising natural resource limits, the earth's 'carrying capacity' relative to the increasing human population, and so forth. How can the creation of an urban commons provide a solution to such challenges? The answer to this question stems from the basic claim advanced in this book: climate change is in essence a political not a natural phenomenon. Because neoliberalism espouses a neo-Malthusian concept of inevitable resource scarcity, all ecological issues are presented in terms of natural limits. In addition, as shown, neoliberalism facilitates corporate class power by dissolving working-class consciousness into the logic of individual self-investment. For all the talk of relentless competition, while the follies of espousing working-class social solidarity are vigorously insisted upon (through such terms as the 'politics of envy', 'class war', and so forth), such solidarity among the corporate class is assiduously cultivated and institutionalised.

Just as Marx revealed in the context of nineteenth-century capitalism, working-class consciousness has always to overturn an initial internalisation of ruling class ideology. In terms of contemporary neoliberal governance, this ideology manifests itself in a simple interdiction: the public may not make its appearance in public. The initial, apparently benign rationale for this restriction is simple: you have no need to appear, given that your existence as 'the people' is presupposed by all public institutions and authorities. In cases of crisis, where this argument about the superfluity of public appearance proves insufficiently strong, a more openly authoritarian line emerges: you will only endanger your own self-interest by your appearance in public and assembly in the commons will create a preordained tragedy compromising public safety and well-being. This transparently paternalistic, anti-democratic logic was on clear display when Occupy encampments were removed. Here the key political issue at stake was the neo-liberal state's determination not to have a commons established through semi-permanent public usage. In other words, this was an overt act of neoliberal state power to prevent the creation of an urban militant particularism.

It is here where we find the linkage between the urban commons and environmental political struggle. Creation of the urban commons would demonstrate that there is no given necessity for the state to assume a neoliberal configuration. It can be converted, admittedly with protracted effort and deliberate organisation, into a genuinely democratic political agency, expressing the will of the people. We need not fear the shadow of totalitarianism here either. For the commons qualifies the power of the state, insofar as the state legally protects usages of land that are defined in an open way by diverse social groupings. Here is where the promise of environmental localism can also find its place within a genuinely progressive environmental politics. Such politics hinges not so much on human agents whose power is essentially independent of material relations to non-humans. The power vested in place through custom is in fact all about a politics of process and relatedness. This is not a rerun of Rousseau's political conception of 'the general will'. Following Latour's thinking, the envisaged politics is instead a meeting place of humans and non-humans, where the mutually enhancing integration of human life within a greater ecological context is the primary goal. Thus, 'the more the merrier' comes to denote a situation where the festive commons extends beyond the modern, exclusively human political community. After all, in a globally urban condition the traditional oppositions between country and city, authentic and inauthentic, natural and artificial, must be passed beyond in forging a credible politics of democratic, associated production.

The idea of a new urban commons sketched out above may seem like an unfortunate and belated relapse into political utopianism given the sobering account of neoliberal environmentalism offered in the earlier chapters of this book. To avoid this pitfall, the sense of effective opposition inherent in the idea of a commons must be stressed. As stated in Chapter 2, neoliberalism can be most readily understood as a process of gradually subordinating all social relations to the logic of market relations. Historically cities have, of course, always been sites of commerce. For Lefevbre (2003 [1970]), as we saw, cities are also fundamentally sites of conflict. In Rose's (1986) legal historical account, this struggle is essentially about places held over for genuine public use. As she puts it: 'Perhaps the chief lesson from the nineteenth-century doctrines of "inherently public property," then, is that while we may change our minds about which activities are socialising, we always accept that the public requires access to some physical locations for some of these activities' (Rose 1986: 781). Neoliberalism, as argued, is better captured as a dynamic process involving a certain paradigm of material (including ecological) development than as some homogeneous bureaucratic rationality. Neoliberalisation realises itself in spatial terms as the gradual marginalisation and eventual elimination of places protected from monetised commerce. In other words, neoliberalism aims at the liquidation of all 'inherently public property'.

Public education, health and housing are obvious examples of key commons whose existence is threatened by neoliberal governance. Those countries where neoliberal reorganisation is most advanced are already in a situation where these public provisions have been in part or wholly destroyed. Even where they survive, a perpetual penumbra of looming crisis overshadows them and makes their long-term survival questionable. Neoliberal governance views all these provisions ultimately as spheres of individual risk management and so ideally navigated by the individual rather than provided en masse through state-run mechanisms. The curious thing, however, is that for all the talk of financial crisis, liberal democratic governments still preside over huge budgets. The neoliberal 'revolution' in areas of public provision seldom amounts to an effort to defund the state, but rather to use state revenues to facilitate aggressively the market alternative. Accordingly, we see privately managed and publicly funded ('charter', 'academy', and so forth) schools being founded at breakneck speed (see Ravitch 2010), public housing sold off, and the piecemeal privatisation of public health systems.

As an antidote to this, the establishment of a new urban commons stands for the more general claim that wealth generated by the majority of citizens must in large part be placed at the disposal of those citizens. To this must be wedded an unwavering environmental perspective based on abundance rather than scarcity. The simple article of faith must be something like this: there is always enough to balance the demands of human satisfaction and ecological well-being. Scarcity only arises because capitalist production works by severing the connection between worker productivity and worker welfare. The energy circulating within human society must be allowed to create a virtuous homeostasis rather than the vicious metastasis exemplified by periodic capitalist crisis.

MARX'S POLITICAL ECOLOGY

The new urban commons, just like its traditional predecessor, can only exist with reference to a legitimising power. That power is the state. I have stressed throughout this analysis that environmental activism is naive where it appeals to localisms and regionalisms that largely bypass the nation state. Despite appearances to the contrary, neoliberal capitalism is not characterised by shrinking state agency, but instead involves a certain style of deploying state power. This is highly visible in the case of neoliberal urbanism, which works by using resources of municipal government to incentivise corporate interventions within the urban environment. Where it does appeal to the state contemporary environmentalism generally focuses on policy change to tighten regulation. While such changes have historically represented important environmental progress, they do not address the minority ownership of production that underlies capitalist environmental action. Adherents of environmental localism often harbour a distrust of the nation state as it represents a centre of power distant from the concerns of communities on the ground.

The typical environmental appeal to local community represents, in certain ways, a sanitised and depoliticised version of Williams's militant particularism. In the context of neoliberalism, however, such localism is much more likely to take the form of reactionary 'NIMBYism' ('not in my backyard') than progressive militant particularism. This is because neoliberal governance presents the physical environment as a context for individual entrepreneurialism. In simple terms, my environment is something I should only want to share on the condition of mutual commercial advantage. Unlike a proposed housing development that directly affects only those on and around a limited site, climate change exists at a scale that transcends not only the local or regional community scale but also that of the national territory. Accordingly, as an expression of post-neoliberal political ecology the new urban commons must achieve the seemingly contradictory status of being both concretely localised but of global compass. This might sound like a rehash of the environmental 'act locally, think globally' mantra. But the key point here is what 'think globally' amounts to in terms of political action.

The only credible agents of global political action are in fact nation states. Even transnational corporations are, despite appearances to the contrary, subject to legal frameworks produced and upheld by internationally coordinated state power. Similarly, the political reality of SD is, whatever environmental localism may hold to the contrary, a global affair. Of course, the promise of the SD platform to distribute environmental benefits and burdens evenly across the world can never be kept in the context of neoliberal capitalism. As argued in previous chapters, the connection between capitalism and uneven development is intrinsic and indefeasible. The mechanisms of neoliberal globalisation do not exist in order to promote universal happiness, as the ongoing European 'migration crisis' forcefully illustrates. Just as those fleeing conflict seek in myriad personal ways to even out uneven development, so those currently able to practice sustainable living do so in an intrinsically unsustainable world.

Ecological theory has long advocated a 'systems thinking' approach to environmental issues (see Meadows 2008). This means, among other things, that there is no 'outside' when considering the ecological effects of human activity. Accordingly, the idea of a new urban commons cannot simply signify a collection of human agents within a geographically delimited area. Marx's idea of a 'metabolic interaction between man and the earth' can point the way here. As John Bellamy Foster (2010) has demonstrated, it is a mistake to view Marx's historical materialism as founded on a Promethean attitude advocating the domination of nature. One of Marx's most direct rejections of this attitude is to be found in his late text (from 1875) *Critique of the Gotha Program*. Here Marx roundly dismisses the idea that all social value is created by human labour alone: 'Labour is *not the source* of all wealth. *Nature* is just as much the source of use values (and it is surely of such that material wealth consists!) as labour, which itself is only the manifestation of a force of nature, human labour power' (Sitton 2010: 150). To this general proposition John Bellamy Foster (2010) adds the following pertinent qualification: 'The chief source of ecological destruction under capitalism, Marx and Engels argued, was the extreme antagonism of town and country, a characteristic of capitalist organisation as fundamental to the system as the division between capitalist and labourer' (Foster 2010: 234).

Images of ecologically disastrous Soviet-era industrial development have allowed for the idea to arise that environmental consciousness and policy are exclusively products of western liberal democracies and their fostering of the market economy. The underlying message of the SD approach is clear: you may think the environment is suffering under neoliberal capitalism, but the consequences of abandoning neoliberalism would bring unimaginably worse ecological consequences. This assumption constitutes another aspect of the post-ideological triumphalism through which neoliberalism has sustained itself over the last three decades. Marx's analysis of capitalism in the nineteenth century helps to undo this complacency by showing how the forces of global capital gave rise to hitherto unknown intensities of resource extraction and circulation. As Foster (2010) makes clear, for Marx the domination of the country by the city was a key index of the capitalist exploitation of nature. The development of synthetic fertilisers along with the importation of such naturally occurring substances as guano from South America allowed mid-nineteenth-century western farming to keep agricultural lands productive by exploiting other geographical regions.

Just as the town/country antagonism could be seen by Marx in terms of colonial removal of natural resources from poorer to richer nation states, so too, at a smaller scale, a city's domination of its rural hinterland was evident. Much of the popular desire to restore and foster local food systems is a largely unconscious will to repair regional rural/urban divides. But these developments are, as is typical of neoliberal environmentalism, largely restricted to patterns of consumption. Accordingly, we are increasingly confronted with a contradiction at the heart of food localism: as demand for local products goes up, the difficulties of local food producers to earn a decent living grow. This is due, in part, to competition resulting in higher land prices. More crucially, however, it stems from the basic fact that the closer to an urban centre, and so to most consumers, an agricultural site is, the greater the discrepancy in land value between residential and retail as opposed to agricultural uses will be. Systematic urban farming under neoliberal capitalism is thus impossible. Unless it adopts a radically different mode of bureaucratic intervention, efforts by metropolitan government to facilitate urban farming will remain marginal and tokenistic. The naive belief that sufficient metropolitan demand will ensure that supply of local produce will keep pace ignores the basic drive of rent extraction that characterises neoliberal urban development.

In this respect earlier efforts such as the 'back to the land' movement of the 1960s and 1970s were more promising (see Agnew 2004). For here, as also in earlier efforts during the Great Depression era (see Nearing and Nearing 1970), there was the basic realisation that ecologically benign ways of life were simply not possible on the terms set by the market economy. Of course, islands of subsistence farming within the context of the world's leading capitalist nation were never going to offer a model that could be extended to the general population without further fundamental social and political changes. Nevertheless, the tens of thousands who did take to the land in the 1970s provided a powerful counter-model to the hegemonic consumerism. In contrast, today's enthusiasm for food localism is conspicuous in its conviction that the virtues of self-sufficiency can be achieved largely on the terms set by neoliberal governance. Until this faith is thoroughly shattered, the desire to eat local will remain yet another fetishised marker of neoliberal environmental consumerism.

The creation of industrialised agriculture, already underway when Marx and Engels were writing in the mid-nineteenth century, is a crucial aspect of capitalism's management of the urban/ rural antagonism. The rapid urbanisation in early industrialising nations such as Britain meant that a wide array of natural limits on labour (seasonality, restriction to daylight hours, and so forth) ceased to apply. Just as mechanisation overcame the conditions of the human body, so too the factory represented an environment where there were no obvious reasons beyond basic physical survival to limit productivity. Hence the early trade union struggles to restrict hours and days of the typical working week. But the same logic of maximised productivity was, of course, also applied to agriculture. The invention of methods for producing synthetic fertiliser in the 1840s by the German chemist Justus von Liebig gradually rendered redundant practical knowledge relating to soil fertility that was passed down through generations of farming families. Liebig's method, along with extraction of naturally occurring fertiliser from colonised territories, essentially removed traditional needs to restrict agricultural usage of the land.

THE NEOLIBERAL FOOD ECONOMY

Just as urban planning in the city, global agriculture reveals the paradoxes of neoliberal governance. In both cases, it is vital not to accept the prevalent wisdom that neoliberalism is all about a diminished role of the state. The interventionism of the neoliberal state is arguably all the more visible in the case of agriculture, thanks to the lavish subsidies granted by the European Union's Common Agricultural Policy (CAP) and the United States Department of Agriculture (USDA) federal Farm Bill. While both sources of state subvention predate the ascendance of neoliberal governance in the 1980s, they have been gradually moulded to the new political paradigm. This paradigm, as we have seen elsewhere, largely presents itself in security terms. 'Food security' is thus the watchword of neoliberal agriculture. Yet such security has little if anything to do with a nation's concerns to feed its own people. This is also true of the apparent beneficence of international 'food aid' lavished on poorer nations by the agricultural powerhouses. In fact, the central paradox of food production under neoliberal capitalism is that it is not ultimately driven by the purpose of feeding people. Once this is realised, it is readily understood that the real social challenges of food production will not be solved by simple increases in yield.

The actual and estimated figures for US state subsidisation for 'foreign agricultural services' (that is, state aid for food exports) between 2013 and 2015 are around $2 trillion. The USDA rationale for this support is as follows:

> Agricultural exports make a critical contribution to the prosperity of local and regional economies across rural America through increased sales and higher commodity prices. Every $1 billion worth of agricultural exports supports an estimated 6,600 jobs and $1.3 billion in economic activity. Because of this important role, the Department is working to reduce trade barriers and develop new markets throughout the world. (USDA 2015a: 14)

Given the putative distaste the neoliberal outlook harbours for this sort of trade-distorting action on the part of the state, the real reasons for this situation have to be looked for elsewhere. The neoliberal state generally regards as anathema direct state subvention of production that cannot be profitable on free market terms (see Graffy 2012). While neoliberal politicians do lip service to saving and creating jobs, the underlying rationale is one of individual risk-taking on the part of workers who should not expect the rest of society to part-finance their employment.

The key social-political fact here is the extraordinary consolidation in US agriculture. According to the USDA, 'four crops (corn, hay, soybeans, and wheat) accounted for over 83 percent of harvested crop acres in 2007. Developments in these few crops drive national trends in midpoint acreages for all cropland combined' (MacDonald et al. 2013: 9). In other words, the development of US agriculture and hence the focus of subsidisation both of production and consumption relates to these big four commodity crops. The USDA estimates that the average farm size today is 1100 acres, though the vast majority of commodity crops are grown on farms much larger (MacDonald et al. 2013: 4). The net cash income for the whole sector ranged between $94 and $126 billion between 2011 and 2014, a large portion of that income in effect coming from state revenue. Farm net income is set to fall sharply for 2015 (around 40 per cent overall) once the figures are in (USDA 2015b).

Both the United States and the European Union (EU) have been subject to repeated international criticism for their subsidising policies as violations of free trade principles (see Clapp 2006). This criticism has come in particular from the Cairns Group of agricultural producers, which includes such agricultural heavyweights as Australia, Canada and Brazil. This group has repeatedly called for the ending of agricultural subsidies since the initial WTO GATT negotiations in the mid-1980s. While the EU has gradually shifted the focus of much subsidisation to various practices of ecological stewardship of agricultural land, the United States has continued to focus on direct support of commodity crop production, exports, and food stamp programmes. Again, the key question for us here is how this situation maintains itself given that the logic of neoliberal economic discipline is more generally intolerant of this type of state action.

Further examination of neoliberal agricultural practice would take us too far from the goal of this final chapter, which is to outline a counter-neoliberal paradigm of political environmental action. What can be readily taken away from this brief analysis, however, is the clear tendency to structure food systems for the sake of profit extraction and the projection of neoliberal state power. Commodity agriculture, just like energy resources, has been subject to the typical neoliberal security configuration. In other words, controlling flows of food is first and foremost a powerful factor in projecting international power. As climate patterns become less predictable, predicting crop yields will become more difficult. As we saw in relation to Welzer's analysis of climate wars, unpredictable patterns of food production are likely to become a more overt factor in the international neoliberal landscape.

All this is more or less ignored by the supporters of food localism, who make the moral rather than political argument that it is preferable to get back in touch with your food. The reality is, however, that the agricultural sector in the wealthiest economies is far from resembling anything like the commons discussed in this section. Most consumers in wealthy liberal democracies who utilise their apparent consumer power by buying 'sustainable' food products are primarily motivated by personal health considerations. Hence, the price premium attached to organic or local produce is tolerated as a canny investment in personal well-being. The conditions of farm labour or the patterns of profit distribution are mostly secondary. Overcoming the SD framework when it comes to food systems, however, requires the creation of a commons analogous to efforts to challenge intensified rent extraction in the context of urban gentrification.

In different ways, the creation of an agricultural commons may be considered both easier and harder than the creation of an urban commons. Easier, insofar as the traditional pre-industrial commons offers a concrete historical model at a time when the majority of the population were still working the land; and harder, because industrialised agricultural takes place within a large, sparsely populated rural landscape. In the United States as in the European Union, this difficulty is greatly exacerbated by the itinerant and transitory nature of the agricultural workforce (see Guthman 2004). It is striking how the richest countries in the world only manage to maintain agricultural production thanks to a largely informal migrant workforce that enjoys few if any of the more usual worker protections. While 'fair trade' labelling commonly confronts consumers on products from developing nations, 'fair work' labelling for domestic food products is rarely a part of the green consumption conversation. A radical critique of the neoliberal SD paradigm has to ask why this is the case.

DIGESTING NEOLIBERALISM

The condition of agricultural land under neoliberalism thus mirrors the fate of the average worker. In the one case a certain portion of land is placed under permanent stress to produce ever-higher yields regardless of the negative effects on the surrounding environment. In the other case, the individual worker is obliged to submit to cycles of productivity intensification that gradually undermine all social conditions of genuine flourishing. Useful natural resources are recklessly used up, just as the implacable forces of market exploitation apply to the worker without necessary reference to social happiness. Foster (1999) cites the following passage from *Capital* to demonstrate how the exploitation of land and worker go hand in hand under global capitalism:

> All progress in capitalist agriculture is a progress in the art, not only of robbing the worker, but of robbing the soil; all progress in increasing the fertility of the soil for a given time is a progress toward ruining the more long-lasting sources of that fertility . . . Capitalist production, therefore, only develops the techniques and the degree of combination of the social process of production by simultaneously undermining the original sources of all wealth – the soil and the worker. (Foster 1999: 379)

Critics such as Arendt who insist that Marx subscribed to a straightforward version of the labour theory of value did not pay attention to this ecological dimension of his critique of capitalism. The key point here, to repeat, is the profound homology between the exploitation of nature (the main concern of environmentalism) and the exploitation of the worker (the chief worry of socialism). The ideological orthodoxy of neoliberal governance over the last three decades is that socialism as a political movement is dead and cannot be revived. Whatever our environmental challenges, it is urged, the solution to them cannot be found by a return to the tradition of socialist economic planning. That experiment has been run and proven to be a dehumanising and often terrifying failure. Neoliberal capitalism, no matter its shortcomings, is the only economic-political dancing partner that can follow the rhythms of a truly democratic, open society.

Climate change points to a contradiction of contemporary capitalism of such complexity and scope that the usual assurances about innovative technological capacity fall well short of what is needed. Among environmental progressives there is mounting disbelief that corporate and political leaders can fail to act in the face of the amassed evidence of climate change. I have argued throughout this book that this surprise is naive. Climate change is not an unforeseen negative consequence of neoliberal governance. Instead, it is a state of affairs perfectly in line with the economic drivers intrinsic to such governance. Subordinating political action in general to market forces means that popular environmental self-determination is not to be tolerated. The manufactured and enforced passivity of the average citizen across contemporary liberal democracies ensures that the only legitimate paradigm for collective environmental action is consumption. Controlling the production of commodities is no more the affair of the democratic collective than the conditions of work are to be determined by the average worker.

Neoliberal governance is, therefore, the direct reverse of the political situation demanded by the socialist workers' movement in the nineteenth century. The generations-long process whereby formal political rights such as enfranchisement granted in the nineteenth and early twentieth century led to material political rights such as workplace self-determination (largely through the trade union movement) has been largely reversed under neoliberalism. This, more than increasing wealth inequality, is the real social harm of neoliberalism. Given this erosion of democratised production it is either naive or disingenuous to argue that systemic environmental problems can be tackled, even in part, by ecologically sensitive consumption. Neoliberalism's enforcement of 'supply-side economics' is engineered to ensure that, where formal political self-determination is conceded, collective material self-determination will be presented as an unnecessary demand. The (globally speaking) relative wealth available for the average citizen across liberal democracies is deemed proof enough that neoliberalism is working for the good of all. But the promise of socialism is quite different, namely one of collective ownership of production. Minority ownership of all land and hence all production, rather than unequal wealth distribution, is the fundamental political pathology of neoliberal governance. Marx's comment on this point, again cited by Foster (1999), anticipates by over a century Brundtland's celebrated definition of sustainable development:

> From the standpoint of a higher socio-economic formation, the private property of particular individuals in the earth will appear just as absurd as the private property of one man in other men. Even an entire society, a nation, or all simultaneously existing societies taken together, are not owners of the earth, they are simply its possessors, its beneficiaries, and have to bequeath it in an improved state to succeeding generations as *boni patres familias* [good heads of the household]. (Foster 1999: 384–5)

If, as noted earlier, for Marx human labour power is ultimately derived from nature, then all wealth creation is a function of natural systems and processes (the basic premise of contemporary ecological economics). If, further, we follow Lefebvre and Merrifield in viewing the condition of urbanisation as having entered a truly global phase of development, then progressive political ecology must perforce think beyond any urban/rural opposition. Credible environmental politics has clearly passed beyond the earlier paradigm of wilderness protection. The problem of climate change, above all, has made this strikingly evident. Conserving pockets of untrammelled nature will do little to prevent the system-wide ecological disturbances entailed.

The call to create or recreate the urban commons is thus, in reality, a universalist proposition that says nothing about the geographical location of the commons. There is, as Occupy Wall Street most obviously demonstrated, a certain tactical advantage in choosing an iconic neoliberal site in which to install the commons. After all, having to capture media attention is a key concern of activists the world over. But it is important not to lose track of the ultimate political function of the commons: namely, to institute and maintain for customary use a network of public places. To ensure that such efforts do not degenerate into a politics of mere spectacle the emphasis of the struggle must be to secure the commons as a legally protected place given over to customary public use. As such sites proliferate the neoliberal push to dissolve public resources into contexts of zero-sum entrepreneurial competition will be faced with concrete opposition. In the context of this struggle it should be remembered that there are in fact many existing commons (schools, libraries and so forth) that exist but require the habitual presence of the public to ensure their ongoing vitality.

Thus, while we must seek to digest neoliberalism in order to pass to something beyond it, it is also important to recall that neoliberal governance has yet to dissolve and absorb many elements of everyday life forged in earlier stages of modern democratic development. By 'digesting' neoliberalism I mean an effort to convert the current neoliberal conjuncture into a (potentially useful) waste product. Looking to reinvention of the commons as a means to pass (beyond) neoliberal governance is partly about bringing the wastage of this social-economic order into public view. As suggested, political good sense tends to view the exposure of the masses as in some way obscene. We can look at this in different ways. On the one hand, fear of the masses in public space might be considered analogous to the offence of displaying the impure. Those who wish democracy to be essentially restricted to well-presented politicians meeting behind closed doors and making occasional, well-orchestrated public appearances, would prefer that 'the people' in all their disturbing heterogeneity not appear.

On the other hand, we could interpret this desire to keep the people hidden as driven by an unconscious impulse analogous to keeping a religious icon away from direct view. In this case, the 'holy' public that ideally directs democratic governance is profaned when put on general view. Most likely, the proscription on the public appearance of the demos is caught up in an irresolvable fluctuation between the pure and the impure. While popular mass media prefer to diagnose political disenchantment in terms of moral disapproval at the corrupt nature of the political class, I believe the real causes lie in the strictly enforced policing of popular self-determination. Thus, 'the people' are disengaged from democratic politics, as there are in truth no real political tasks for them to perform.

The interrupted evolution of democracy mostly comes down to reproducing a condition of moral-political minority with regard to the enfranchised public. It seems curious, to say the least, that those empowered to determine their political representatives are deemed incapable of direct political self-determination. At least a classical nineteenth-century liberal such as John Stuart Mill was consistent when he made adequate education and moral cultivation a precondition of all valid popular political praxis. Somewhat surprisingly, in his posthumously published *Chapters on Socialism*, Mill (1989 [1879]) was willing to concede that certain models of socialism (he favoured Charles Fourier's vision of small, autonomous socialist communities) could indeed offer a superior model for maximising social happiness relative to the prevailing market economy. Mill's principal worry on this count related to psychological motivation to work hard and honestly. Any socialist arrangement that distributed rewards based entirely on perceived needs rather than displayed merits would, he felt, lead to a harmful falling off of efficient production. On balance, therefore, he thinks the socialist transition would only be possible when the technical and moral education of all members of the community has reached a certain level. In his own words:

> The result of our review of the various difficulties of Socialism has led us to the conclusion that the various schemes for managing the productive resources of the country by public instead of private agency have a case for trial, and some of them may eventually establish their claims to preference over the existing order of things, but that they are at present workable only by the élite of mankind, and have yet to prove their power of training mankind at large to the state of improvement which they presuppose. (Mill 1989: 273)

Mill's progressive liberalism is a far cry from the present-day cynicism of neoliberal capitalism. The 'post-historical' condition of neoliberal governance rules out the kind of progressive optimism that characterised liberalism as a nineteenth-century political force. Historical progression in our times has been reduced to the circuits of accelerated commodity obsolescence and the fantasies of apocalyptic 'transhumanism'. In all this, the spectre of climate change reminds us that, for all our technical advance, 'nature' has not been abolished. Nature's 'end', in the sense McKibben and other environmentalists have construed this notion over the last few decades, is really a manner of experiencing nature through a catastrophic collective consciousness.

CONCLUSION

This book has been an attempt to interpret, behind this environmental construction, the way in which neoliberal governance produces nature. Climate change is, at base, a political condition that calls for political analysis and, above all, alternative modes of political praxis. In ecological terms, climate change points to material wastage of such a degree that there cannot be recirculation and absorption without systemic destabilisation. Within the neoliberal order destabilising change translates into entrepreneurial opportunity. Insofar as growing evidence of climate change has produced a political response on the part of leading neoliberal states, a new front in the deployment of military resources to safeguard national security is being opened up.

As environmentalists push harder on the natural catastrophe narrative, the reactive neoliberal state and the corporate ruling class are gearing up their own response to imminent crisis or collapse. This unnerving convergence between progressive environmental activists and reactionary government and business should in itself be taken as an indication that something is greatly amiss. Modern environmentalism needs to overcome its ruling discourse of resource scarcity and limits and return to the experimental forms of socialism that flowered in such abundance in the nineteenth century. Analogously, popular politics across liberal democracies needs to shift its emphasis away from the obsessive focus on wealth distribution and return to the more fundamental questions of the communist and socialist heritage that ultimately relate to material self-governance. In other words, as I have argued throughout this book, we need to focus unwaveringly on the massive and inevitable human and ecological wastage produced by the neoliberal order and slowly return to a truly 'sustainable' organisation of the material sources of life. This is, in the end, the only real problem of political ecology. In this light, climate change should not be apprehended as a sublime object portending the end of the world as we know it, but rather as an invitation to recommence work on creating a society of freely associated producers.

Let us return to the theme of catastrophe with which our enquiry began. Climate change is being predominantly and increasingly presented as imminent environmental collapse. Accordingly, political efforts to reverse or prepare for the effects of climate change must occur within years to prevent massive disruption to the earth's natural systems. I have argued that this appeal is implausible and will likely be extremely counter-effective on the level of political praxis because there is insufficient critical analysis of what is driving climate change. The root cause of climate change is not the nefarious actions and vicious character of powerful CEOs and members of the political class. Nor it is the result of unethical consumer choices in more affluent societies. Instead, both sets of actors are made possible by the political logic operative in the form of neoliberal governance. Such governance constitutes a social system focusing on the core value of entrepreneurialism. The neoliberal paradigm over the last forty years has proven sufficiently powerful to undermine democratic institutions and political common sense across wealthy liberal democracies. Conversely, undoing the neoliberal order will not be a matter of a quick, concerted push on the part of a well-organised group or network of environmental activists. Such activists by and large present climate change in the language of natural limits and see the challenges sufficiently stated using climate science, even where appeals to alternative patterns of political governance are seen as part of the answer.

I have argued throughout this book, by contrast, that it is essential to see climate change as a radically political phenomenon. As such it is strictly speaking irrelevant whether climate change is an intended or unintended consequence of neoliberal governance. As a general horizon of social understanding, neoliberalism is the kind of thing Marx referred to as an ideology and the early Foucault called an episteme: a historically embedded, largely unconscious horizon of collective understanding and action. As argued, neoliberalism's negative environmental impact is intimately tied up with its intolerance for all forms of popular collective ownership. That is why the first task for a politically enlightened environmentalism would be the creation of a network of numerous, high-profile commons. Thus, the Occupy movement was on the right track. But the demand for a conspicuous place of popular appearance in the urban centre must then reverberate into calls for workplace control by workers; calls to get public education back into the hands of the people; food back under the control of local producers, and so forth. Democracy's evolution has essentially been stalled by the growth of the consumer society and neoliberal governance. The environmental effects of neoliberal governance up to and including climate change can only be tackled by a generations-long, drawn out effort across wealthy liberal democracies to place the material environment of production under popular control. Nothing less will constitute a plausible, effective and sufficiently radical response to the political problem of climate change. At this point in time there can little doubt that almost all the work needed to mount this response remains to be done.

Bibliography

Ackroyd, Peter (2000), *London: The Biography*, New York: Doubleday.

Adorno, Theodore and Max Horkheimer (1997 [1944]), *Dialectic of Enlightenment*, London: Verso.

Agnew, Eleanor (2004), *Back from the Land: Why Young Americans Went to Nature in the 1970s and Why They Came Back*, Chicago: Ivan R. Dee.

Anderson, Benedict (2006 [1983]), *Imagined Communities: Reflections of the Origin and Spread of Nationalism*, London: Verso.

Aristotle (1999), *Metaphysics*, London: Penguin.

Aristotle (2009), *Politics*, Oxford: Oxford University Press.

Arendt, Hannah (1998 [1958]), *The Human Condition*, Chicago: University of Chicago Press.

Benjamin, Walter (2007 [1955]), *Illuminations*, New York: Random House.

Biermann, Frank and Ingrid Boas (2010), 'Preparing for a warmer world: towards a global governance system to protect climate refugees', *Global Environmental Politics*, 10(1): 60–88.

BP (2016), 'Our strategy and sustainability', <http://www.bp.com/en/global/corporate/sustainability/bp-and-sustainability/our-strategy-and-sustainability.html> (last accessed 23 February 2016).

Bronner, Stephen and Douglas Kellner (1989), *Critical Theory and Society: A Reader*, New York: Routledge.

Bullard, Robert (ed.) (1993), *Confronting Environmental Racism: Voices from the Grassroots*, Boston: South End Press.

Bullard, Robert (ed.) (2000 [1990]), *Dumping in Dixie: Race, Class, and Environmental Quality*, Boulder, CO: Westview Press.

Bullard, Robert (ed.) (2003), *Just Sustainabilities: Development in an Unequal World*, Cambridge, MA: MIT Press.

Burke, Edmund (2015 [1757]), *A Philosophical Enquiry Into the Origin of Our Ideas of the Sublime and Beautiful*, Oxford: Oxford University Press.

Canetti, Elias (1984 [1960]), *Crowds and Power*, New York: Farrar, Straus and Giroux.

Cederström, Carl and André Spicer (2015), *The Wellness Syndrome*, Cambridge: Polity.

Chomsky, Noam (2005), 'Containing the threat of democracy', in *Chomsky on Anarchism*, Oakland: AK Press, pp. 153–75.

Clapp, Jennifer (2006), 'WTO agriculture negotiations: implications for the global South', *Third World Quarterly*, 27: 563–77.

Clapp, Jennifer (2012), *Food*, Cambridge: Polity Press.

Colgan, J. D. (2014), 'Oil, domestic politics, and international conflict', *Energy Research and Social Science*, 1: 198–205.

Costanza, Robert (1992), *Ecological Economics: The Science and Management of Sustainability*, New York: Columbia University Press.

Cotula, Lorenzo (2013), *The Great African Land Grab?: Agricultural Investments and the Global Food System*, London: Zed Books.

Cronon, William (1995), 'The trouble with wilderness; or, getting back to the wrong nature', in W. Cronon (ed.), *Uncommon Ground: Rethinking the Human Place in Nature*, New York: W. W. Norton, pp. 69–90.

Davis, Mike (2005), *Planet of Slums*, London: Verso.

Davis, Mike (2010), 'Who will build the ark', *New Left Review*, 61: 29–46.

Davutoğlu, Ayten (2013), 'Two different poverty reduction approaches: neoliberal market based microfinance versus social rights defender basic income', *International Journal of Social Inquiry*, 6: 39–47.

Debord, Guy (1994 [1967]), *The Society of the Spectacle*, New York: Zone.

De Castro, Fabio, Barbara Hogenboom and Michiel Baud (2016), *Environmental Governance in Latin America*, New York: Palgrave Macmillan.

Diamond, Jared (2005), *Collapse: How Societies Choose to Fail or Succeed*, New York: Penguin.

Elliott, Brian (2010), *Constructing Community: Configurations of the Social in Twentieth-Century Urbanism and Philosophy*, Lanham, MD: Lexington.

Engels, Friedrich (1988 [1872]), 'The housing question', in *Karl Marx and Friedrich Engels: Collected Works*, New York: Internationalist Publishers, 23: 215–91.

Enwegbara, Basil (2001), 'Toxic colonialism: Lawrence Summers and let Africans eat pollution', *The Tech*, MIT, < http://tech.mit.edu/V121/N16/col16guest.16c.html> (last accessed 4 March 2016).

European Commission (2016), Agriculture and Rural Development, 'Agri-food trade statistical factsheet', <http://ec.europa.eu/agriculture/trade-analysis/statistics/outside-eu/regions/agrifood-extra-eu-28_en.pdf> (last accessed 24 March 2016).

Exxon Mobil (2015), 'Sustainability in motion', <https://exxonmobil.com/lubes/sustainability.aspx> (last accessed 28 August 2015).

Friedman, Milton, (2002 [1962]), Capitalism and Freedom, Chicago: University of Chicago Press.

Friedman, Milton and Rose Friedman (1990 [1980]), Free to Choose: A Personal Statement, Orlando: Harcourt.

Foster, John Bellamy (1999), 'Marx's theory of metabolic rift: classical foundations for environmental sociology', American Journal of Sociology, 105(2): 366–405.

Foster, John Bellamy (2010 [1995]), 'Marx and the environment', in John Sitton (ed.), Marx Today: Selected Works and Recent Debates, New York: Palgrave Macmillan, pp. 229–40.

Foster, John Bellamy (2013), 'Marx and the rift in the universal metabolism of nature', Monthly Review, 65(7).

Foucault, Michel (2008), The Birth of Biopolitics. Lectures at the Collège de France 1978–1979, New York: Picador.

Fukuyama, Francis (1992), The End of History and the Last Man, New York: Free Press.

Funk, Mackenzie (2015), Windfall: The Booming Business of Global Warming, New York: Penguin.

Gates, Charles (2011), Ancient Cities: The Archaeology of Urban Life in the Ancient Near East and Egypt, Greece and Rome, London: Routledge.

Gentleman, Amelia (2014), 'Food bank Britain: can MPs agree on the causes of poverty in the UK?', The Guardian, 4 July, <http://www.theguardian.com/society/2014/jul/04/food-bank-britain-can-mps-agree-causes-uk-poverty> (last accessed 11 January 2016).

Giddens, Anthony (1985), The Nation-State and Violence, Cambridge: Polity.

Giddens, Anthony (1998), The Third Way: The Renewal of Social Democracy, Cambridge: Polity.

Giddens, Anthony (2011), The Politics of Climate Change, Cambridge: Polity.

Graffy, Elizabeth (2012), 'Agrarian ideals, sustainability ethics, and US policy', Journal of Agricultural Environmental Ethics, 25: 503–28.

Graham, Stephen (2010), Cities under Siege: The New Military Urbanism, London: Verso.

Guthman, Julie (2004), *Agrarian Dreams: The Paradox of Organic Farming in California*, Oakland, CA: University of California Press.

Haas, Tigran (2008), *New Urbanism and Beyond: Designing Cities for the Future*, New York: Rizzoli.

Harvey, David (1996), *Justice, Nature and the Geography of Difference*, Oxford: Blackwell.

Harvey, David (1998), 'What's green and makes the environment go round?' in F. Jameson and M. Miyoshi (eds), *The Cultures of Globalization*, Durham, NC: Duke University Press, pp. 327–55.

Harvey, David (2000), *Spaces of Hope*, Berkeley and Los Angeles: University of California Press.

Harvey, David (2005), *A Brief History of Neoliberalism*, Oxford: Oxford University Press.

Harvey, David (2006 [1982]), *The Limits to Capital*, Oxford: Blackwell.

Harvey, David (2011), *The Enigma of Capital*, Oxford: Oxford University Press.

Harvey, David (2012), *Rebel Cities: From the Right to the City to the Urban Revolution*, London: Verso.

Hatherley, Owen (2016), *The Ministry of Nostalgia*, London: Verso.

Hayek, Friedrich von (2011 [1960]), *The Constitution of Liberty*, Chicago: University of Chicago Press.

Hoffmann, Andrew (2015), *How Culture Shapes the Climate Change Debate*, Stanford, CA: Stanford University Press.

Hopkins, Bob (2013), *The Power of Just Doing Stuff: How Local Action Can Change the World*, Cambridge: UIT Cambridge.

IPCC (2014), *Climate Change 2014: Synthesis Report*, Contribution of Working Groups I, II and III to the Fifth Assessment Report of the Intergovernmental Panel on Climate Change, <https://www.ipcc.ch/pdf/assessment-report/ar5/syr/SYR_AR5_FINAL_full.pdf> (last accessed 24 March 2016).

Jaffee, Daniel (2014), *Brewing Justice: Fair Trade Coffee, Sustainability, and Survival*, Berkley and Los Angeles: University of California Press.

Jones, Owen (2011), *Chavs: The Demonization of the Working Class*, London: Verso.

Klein, Naomi (2007), *The Shock Doctrine: The Rise of Disaster Capitalism*, New York: Picador.

Klein, Naomi (2014), *This Changes Everything: Capitalism vs. the Climate*, New York: Basic Books.

Klein, Naomi (2015), 'Change everything or face a global Katrina', <http://www.naomiklein.org/articles/2015/08/change-everything-or-face-global-katrina> (last accessed 24 April 2016).

Knabb, Ken (1995), *Situationist International Anthology*, Berkeley: Bureau of Public Secrets.

Kuhn, Thomas (2012), *The Structure of Scientific Revolutions*, Chicago: University of Chicago Press.

Latour, Bruno (1999), *Pandora's Hope: Essays on the Reality of Science Studies*, Cambridge, MA: Harvard University Press.

Latour, Bruno (2004 [1999]), *Politics of Nature: How to Bring the Sciences into Democracy*, Cambridge, MA: Harvard University Press.

Lefebvre, Henri (1991 [1973]), *The Production of Space*, Oxford: Blackwell.

Lefebvre, Henri (2003 [1970]), *The Urban Revolution*, Minneapolis: University of Minnesota Press.

Lewis, Joanna (2014), 'The rise of renewable energy protectionism: emerging trade conflicts and implications for low carbon development', *Global Environmental Politics*, 14(4): 10–25.

Lyotard, Jean-François (1988 [1983]), *The Differend: Phases in Dispute*, Minneapolis: Minnesota University Press.

Lyotard, Jean-François (1991 [1988]), *The Inhuman: Reflections on Time*, Cambridge: Polity.

Macalister, Terry (2015), 'Shell abandons Alaska Arctic drilling', *The Guardian*, 28 September, <http://www.theguardian.com/business/2015/sep/28/shell-ceases-alaska-arctic-drilling-exploratory-well-oil-gas-disappoints> (last accessed 19 February 2016).

MacDonald, James, Penni Korb and Robert Hoppe (2013), *Farm Size and the Organization of U.S. Farm Cropping*, United States Department of Agriculture, Economic Research Report Number 152, <http://www.ers.usda.gov/media/1156726/err152.pdf> (last accessed 16 October 2015).

McKibben, Bill (2006 [1989]), *The End of Nature*, New York: Random House.

McKibben, Bill (2007), *Deep Economy: The Wealth of Communities and the Durable Future*, New York: Times Books.

Marcuse, Herbert (1991 [1964]), *One-Dimensional Man: Studies in the Ideology of Advanced Industrial Society*, Boston: Beacon.

Marshall, George (2014), *Don't Even Think About It: Why Our Brains Are Wired to Ignore Climate Change*, New York: Bloomsbury.

Marx, Karl (1994 [1888]), 'On Feuerbach', in Joseph O'Malley (ed.), *Marx: Early Political Writings*, Cambridge: Cambridge University Press.

Mason, Paul (2015), *PostCapitalism: A Guide to our Future*, London: Allen Lane.

Meadows, Donella (2008), *Thinking in Systems: A Primer*, White River Junction, VT: Chelsea Green Publishing.

Meadows, Donella H., Dennis L. Meadows, Jørgen Randers and William H. Behrens III (1972), *The Limits to Growth*, New York: Signet.

Merrifield, Andy (2013), 'The urban question under planetary urbanization', *International Journal of Urban and Regional Research*, 37(3): 909–22.

Mill, John Stuart (1989 [1879]), 'Chapters on Socialism', in Stefan Collini (ed.), *On Liberty and Other Writings*, Cambridge: Cambridge University Press.

Mitchell, Don (2003), *The Right to the City: Social Justice and the Fight for Public Space*, New York: Guilford Press.

Naess, Arne (1973), 'The shallow and the deep, long-range ecology movement: a summary', *Inquiry: An Interdisciplinary Journal of Philosophy and the Social Sciences*, 16: 95–100.

Nearing, Helen and Scott Nearing (1970), *Living the Good Life: How to Live Simply and Sanely in a Troubled World*, New York: Shocken.

Nordhaus, William D. (2007). 'A review of the *Stern Review on the Economics of Climate*', *Journal of Economic Literature*, 45: 686–702.

Norgaard, Kari (2011), *Living in Denial: Climate Change, Emotions, and Everyday Life*, Cambridge, MA: MIT Press.

Nozick, Robert (2013 [1974]), *Anarchy, State, and Utopia*, New York: Basic Books.

Piketty, Thomas (2014 [2013]), *Capital in the Twenty-First Century*, Cambridge, MA: Belknap.

Plato (2008), *Republic*, Oxford: Oxford University Press.

Rancière, Jacques (2014 [2005]), *Hatred of Democracy*, London: Verso

Ravitch, Diane (2010), *The Death and Life of the Great American School System: How Testing and Choice are Undermining Education*, New York: Basic Books.

Rebarber, Ted and Alison Consoletti Zgainer (2014), Survey of America's charter schools, Center for Education Reform, <https://www.edreform.com/wp-content/uploads/2014/02/2014Charter SchoolSurveyFINAL.pdf> (last accessed 4 March 2016).

Rose, Carol (1986), 'The comedy of the commons: commerce, custom, and inherently public property', *Faculty Scholarship Series*, 53(3): 711–81.

Sartre, Jean-Paul (1963), *Search for a Method*, New York: Knopf.

Schumacher, E. F. (1973), *Small is Beautiful*, London: Blond and Briggs Ltd.

Shell (2016), 'What sustainability means at Shell', <http://www.shell.com/sustainability/our-approach/sustainability-at-shell.html> (last accessed 23 February 2016).

Sitton, John (2010), *Marx Today: Selected Works and Recent Debates*, New York: Palgrave Macmillan.

Skinner, B. F. (1976), *About Behaviorism*, New York: Vintage.

Sloterdijk, Peter (2013 [2005]), *In the World Interior of Capital: Towards a Philosophical Theory of Globalization*, Cambridge: Polity.

Smith, Neil (2008 [1984]), *Uneven Development: Nature, Capital, and the Production of Space*, Athens, GA: University of Georgia Press.

Spash, Clive L. (2010), 'The brave new world of carbon trading', *New Political Economy*, 15(2): 169–95.

Srnicek, Nick and Alex Williams (2015), *Inventing the Future: Postcapitalism and a World Without Work*, London: Verso.

Steen-Olsen, Kjartan, Jan Weinzettl, Gemma Cranston and Edgar G. Hertwich (2012), 'Carbon, land, and water footprint accounts for the European Union: consumption, production, and displacements through international trade', *Environmental Science and Technology*, 46(20): 10883–91.

Steffen, Will, Paul J. Crutzen and John R. McNeill (2007), 'The Anthropocene: are humans now overwhelming the great forces of nature?', *AMBIO: A Journal of the Human Environment*, 36(8): 614–21.

Stern, Nicholas (2007), *The Economics of Climate Change*, Cambridge: Cambridge University Press.

Sullivan, Sian (2013), 'Banking nature? The spectacular financialisation of environmental conservation', *Antipode*, 45(1): 198–217.

Turner, R. Kerry, David Pearce and Ian Bateman (1994), *Environmental Economics: An Elementary Introduction*, New York: Harvester Wheatsheaf.

Umejesi, I. and W. Akpan (2013), 'Oil exploration and local opposition in colonial Nigeria: understanding the roots of contemporary state-community conflict in the Niger Delta', *South African Review of Sociology*, 44(1): 111–30.

United Nations (1987), *Report of the World Commission on Environment and Development: Our Common Future* ('The Brundtland Report'), <http://www.un-documents.net/our-common-future.pdf> (last accessed 10 November 2014).

United Nations Development Programme (1994), *Human Development Report*, <http://hdr.undp.org/sites/default/files/reports/255/hdr_1994_en_complete_nostats.pdf> (last accessed 4 March 2016).

United States Department of Agriculture (2015a), *FY 2015 Budget Summary and Annual Performance Plan*, <http://www.obpa.usda.gov/budsum/FY15budsum.pdf> (last accessed 15 October 2015).

United States Department of Agriculture (2015b), *Farm Sector Income Forecast* 2015, <http://www.ers.usda.gov/topics/farm-economy/farm-sector-income-finances/2015-farm-sector-income-forecast.aspx> (last accessed 16 October 2015).

United States Department of Defense (2014), *Climate Change Adaptation Roadmap*, <http://www.acq.osd.mil/ie/download/CCARprint_wForeword_c.pdf> (last accessed 16 September 2015).

United States Department of Labor (Bureau of Labor Statistics) (2014), *Union Members Summary*, <http://www.bls.gov/news.release/union2.nro.htm> (last accessed 15 November 2014).

Vanderheiden, Steve (2005), 'Eco-terrorism or justified resistance? Radical environmentalism and the "War on Terror"', *Politics and Society*, 33(3): 425–47.

Welzer, Harald (2012), *Climate Wars: What People Will be Killed For in the 21st Century*, Cambridge: Polity.

The Whitehouse, 'The National Security Implications of Climate Change', <https://www.whitehouse.gov/sites/default/files/docs/National_Security_Implications_of_Changing_Climate_Final_051915.pdf> (last accessed 20 September 2015).

Williams, Raymond (1973), *The Country and the City*, Oxford: Oxford University Press.

Williams, Raymond (1989), *Resources of Hope*, London: Verso.

Williams, Raymond (2005 [1980]), 'A hundred years of *Culture and Anarchy*', in *Culture and Materialism*, London: Verso, pp. 3–9.

Wintour, Patrick (2015), 'Britain should not take more Middle East refugees, says David Cameron', *The Guardian*, 3 September 2015, <http://www.theguardian.com/world/2015/sep/02/david-cameron-migration-crisis-will-not-be-solved-by-uk-taking-in-more-refugees> (last accessed 4 March 2016).

Wood, Ruth, Paul Gilbert, Maria Sharmina et al. (2011), 'Shale gas: a provisional assessment of climate change and environmental impacts', <http://www.tyndall.ac.uk/sites/default/files/coop_shale_gas_report_final_200111.pdf> (last accessed 10 January 2016).

World Trade Organization (2016), 'Sustainable development', <https://www.wto.org/english/tratop_e/envir_e/sust_dev_e.htm> (last accessed 21 March 2016).

Zalasiewicz, Jan, Mark Williams, Alan Haywood, Michael Ellis (2011), 'The Anthropocene: a new epoch of geological time?', *Philosophical Transactions of the Royal Society of London A: Mathematical, Physical and Engineering Sciences*, 369(1938): 835–41.

Index